Philosophy of Science
Part III

Professor Jeffrey L. Kasser

THE TEACHING COMPANY ®

PUBLISHED BY:

THE TEACHING COMPANY
4151 Lafayette Center Drive, Suite 100
Chantilly, Virginia 20151-1232
1-800-TEACH-12
Fax—703-378-3819
www.teach12.com

Copyright © The Teaching Company Limited Partnership, 2006

Printed in the United States of America

This book is in copyright. All rights reserved.

Without limiting the rights under copyright reserved above,
no part of this publication may be reproduced, stored in
or introduced into a retrieval system, or transmitted,
in any form, or by any means
(electronic, mechanical, photocopying, recording, or otherwise),
without the prior written permission of
The Teaching Company.

ISBN 1-59803-240-2

Jeffrey L. Kasser, Ph.D.

Teaching Assistant Professor, North Carolina State University

Jeff Kasser grew up in southern Georgia and in northwestern Florida. He received his B.A. from Rice University and his M.A. and Ph.D. from the University of Michigan (Ann Arbor). He enjoyed an unusually wide range of teaching opportunities as a graduate student, including teaching philosophy of science to Ph.D. students in Michigan's School of Nursing. Kasser was the first recipient of the John Dewey Award for Excellence in Undergraduate Education, given by the Department of Philosophy at Michigan. While completing his dissertation, he taught (briefly) at Wesleyan University. His first "real" job was at Colby College, where he taught 10 different courses, helped direct the Integrated Studies Program, and received the Charles Bassett Teaching Award in 2003.

Kasser's dissertation concerned Charles S. Peirce's conception of inquiry, and the classical pragmatism of Peirce and William James serves as the focus of much of his research. His essay "Peirce's Supposed Psychologism" won the 1998 essay prize of the Charles S. Peirce Society. He has also published essays on such topics as the ethics of belief and the nature and importance of truth. He is working (all too slowly!) on a number of projects at the intersection of epistemology, philosophy of science, and American pragmatism.

Kasser is married to another philosopher, Katie McShane, so he spends a good bit of time engaged in extracurricular argumentation. When he is not committing philosophy (and sometimes when he is), Kasser enjoys indulging his passion for jazz and blues. He would like to thank the many teachers and colleagues from whom he has learned about teaching philosophy, and he is especially grateful for the instruction in philosophy of science he has received from Baruch Brody, Richard Grandy, James Joyce, Larry Sklar, and Peter Railton. He has also benefited from discussing philosophy of science with Richard Schoonhoven, Daniel Cohen, John Carroll, and Doug Jesseph. His deepest gratitude, of course, goes to Katie McShane.

Table of Contents
Philosophy of Science
Part III

Professor Biography i
Course Scope 1
Lecture Twenty-Five   New Views of Meaning and Reference 4
Lecture Twenty-Six   Scientific Realism 23
Lecture Twenty-Seven   Success, Experience, and Explanation 42
Lecture Twenty-Eight   Realism and Naturalism 60
Lecture Twenty-Nine   Values and Objectivity 78
Lecture Thirty   Probability 96
Lecture Thirty-One   Bayesianism 116
Lecture Thirty-Two   Problems with Bayesianism 135
Lecture Thirty-Three   Entropy and Explanation 154
Lecture Thirty-Four   Species and Reality 172
Lecture Thirty-Five   The Elimination of Persons? 191
Lecture Thirty-Six   Philosophy and Science 210
Timeline 227
Glossary 240
Biographical Notes 252
Bibliography 257

Philosophy of Science

Scope:

With luck, we'll have informed and articulate opinions about philosophy and about science by the end of this course. We can't be terribly clear and rigorous prior to beginning our investigation, so it's good that we don't need to be. All we need is some confidence that there is something about science special enough to make it worth philosophizing about and some confidence that philosophy will have something valuable to tell us about science. The first assumption needs little defense; most of us, most of the time, place a distinctive trust in science. This is evidenced by our attitudes toward technology and by such notions as who counts as an expert witness or commentator. Yet we're at least dimly aware that history shows that many scientific theories (indeed, almost all of them, at least by one standard of counting) have been shown to be mistaken. Though it takes little argument to show that science repays reflection, it takes more to show that philosophy provides the right tools for reflecting on science. Does science need some kind of philosophical grounding? It seems to be doing fairly well without much help from us. At the other extreme, one might well think that science occupies the entire realm of "fact," leaving philosophy with nothing but "values" to think about (such as ethical issues surrounding cloning). Though the place of philosophy in a broadly scientific worldview will be one theme of the course, I offer a preliminary argument in the first lecture for a position between these extremes.

Although plenty of good philosophy of science was done prior to the 20th century, nearly all of today's philosophy of science is carried out in terms of a vocabulary and problematic inherited from logical positivism (also known as logical empiricism). Thus, our course will be, in certain straightforward respects, historical; it's about the rise and (partial, at least) fall of logical empiricism. But we can't proceed purely historically, largely because logical positivism, like most interesting philosophical views, can't easily be understood without frequent pauses for critical assessment. Accordingly, we will work through two stories about the origins, doctrines, and criticisms of the logical empiricist project. The first centers on notions of meaning and evidence and leads from the positivists through the work of Thomas Kuhn to various kinds of social constructivism and postmodernism. The second story begins from the notion of explanation and culminates in versions of naturalism and scientific realism. I freely grant that the separation of these stories is somewhat artificial, but each tale stands tolerably well on its own, and it will prove helpful to look at similar issues from distinct but complementary angles. These narratives are sketched in more detail in what follows.

We begin, not with logical positivism, but with a closely related issue originating in the same place and time, namely, early-20th-century Vienna. Karl Popper's provocative solution to the problem of distinguishing science from pseudoscience, according to which good scientific theories are *not* those that are highly confirmed by observational evidence, provides this starting point. Popper was trying to capture the difference he thought he saw between the work of Albert Einstein, on the one hand, and that of such thinkers as Sigmund Freud, on the other. In this way, his problem also serves to introduce us to the heady cultural mix from which our story begins.

Working our way to the positivists' solution to this problem of demarcation will require us to confront profound issues, raised and explored by John Locke, George Berkeley, and David Hume but made newly urgent by Einstein, about how sensory experience might constitute, enrich, and constrain our conceptual resources. For the positivists, science exhausts the realm of fact-stating discourse; attempts to state extra-scientific facts amount to metaphysical discourse, which is not so much false as meaningless. We watch them struggle to reconcile their empiricism, the doctrine (roughly) that all our evidence for factual claims comes from sense experience, with the idea that scientific theories, with all their references to quarks and similarly unobservable entities, are meaningful and (sometimes) well supported.

Kuhn's historically driven approach to philosophy of science offers an importantly different picture of the enterprise. The logical empiricists took themselves to be explicating the "rational core" of science, which they assumed fit reasonably well with actual scientific practice. Kuhn held that actual scientific work is, in some important sense, much less rational than the positivists realized; it is driven less by data and more by scientists' attachment to their theories than was traditionally thought. Kuhn suggests that science can only be understood "warts and all," and he thereby faces his own fundamental tension: Can an understanding of what is intellectually special about science be reconciled with an understanding of actual scientific practice? Kuhn's successors in sociology and philosophy wrestle (very differently) with this problem.

The laudable empiricism of the positivists also makes it difficult for them to make sense of causation, scientific explanation, laws of nature, and scientific progress. Each of these notions depends on a kind of connection or structure that is not present in experience. The positivists' struggle with these notions provides the occasion for our second narrative, which proceeds through new developments in meaning and toward scientific realism, a view that seems as commonsensical as empiricism but stands in a deep (though perhaps not irresolvable) tension with the latter position. Realism (roughly) asserts that scientific theories can and sometimes do provide an accurate picture of reality, including unobservable reality. Whereas constructivists appeal to the theory-dependence of observation to show that we help constitute reality, realists argue from similar premises to the conclusion that we can track an independent reality. Many realists unabashedly use science to defend science, and we examine the legitimacy of this naturalistic argumentative strategy. A scientific examination of science raises questions about the role of values in the scientific enterprise and how they might contribute to, as well as detract from, scientific decision-making. We close with a survey of contemporary application of probability and statistics to philosophical problems, followed by a sketch of some recent developments in the philosophy of physics, biology, and psychology.

In the last lecture, we finish bringing our two narratives together, and we bring some of our themes to bear on one another. We wrestle with the ways in which science simultaneously demands caution and requires boldness. We explore the tensions among the intellectual virtues internal to science, wonder at its apparent ability to balance these competing virtues, and ask how, if at all, it could do an even better job. And we think about how these lessons can be deployed in extra-scientific contexts. At the end of the day, this will turn out to have been a course in conceptual resource management.

Lecture Twenty-Five
New Views of Meaning and Reference

Scope:

A new philosophical theory of reference and meaning makes it easier to face problems of incommensurability; philosophers can now more readily say that we have a new theory about the same old mass rather than a theory of Einsteinian mass competing with a theory of Newtonian mass. The new theory, for better and for worse, also makes it easier to talk about unobservable reality. In this lecture, we explore this new approach to meaning and reference, along with a new conception of scientific theories that accompany it. Scientific theories are now sometimes conceived in terms of models and analogies, rather than as deductive systems. We also consider some legitimate worries the once-received view poses for the new view.

Outline

I. At this point, we begin bringing our two narratives together by integrating issues of meaning and reference into our recent discussions of explanation and allied notions. We have been tacitly relying on a fairly standard philosophical account of reference, according to which we typically pick things out by correctly describing them.

   A. Meaning and reference are distinct. *Albert Einstein* and *the discoverer of special relativity* co-refer, but they do not have the same meaning. Likewise, *creature with a heart* applies to all the same things as *creature with a kidney*, but they don't mean the same thing.

   B. In a standard understanding, a description such as *the favorite physicist of the logical positivists* must correctly pick out a unique individual (for example, Einstein) in order to refer.

   C. Suppose that, unbeknownst to me, Werner Heisenberg turns out to be the favorite physicist of the logical positivists. In that case, I may think I am using the phrase to refer to Einstein, but I am really referring to Heisenberg.

- **D.** As we have seen, the logical positivists treated meaning and reference as relatively unproblematic for observational terms and as quite problematic for theoretical terms.
- **E.** A common version of this approach does not provide reference for theoretical terms at all; the parts of scientific theory that are not about experience do not directly refer to the world and do not aspire to truth. Talk of quarks just serves to systematize and predict observation.
- **F.** Less stringent empiricists allowed theoretical terms to refer and treated them in the standard way. This is the approach taken by Thomas Kuhn.
    1. For Kuhn, reference is fairly easy to secure, because a term refers only to the world-as-described-by-the-paradigm. Thus, in Kuhn's view, such a term as *phlogiston* refers just as surely as *oxygen* does to something that can cause combustion; both refer to crucial causes of combustion, as identified by their paradigms.
    2. This makes reference too easy to secure. Most philosophers find it much more natural to say that phlogiston never existed, and the term *phlogiston* never referred to anything.
- **G.** On the other hand, the standard view makes reference too hard to secure. If Benjamin Franklin misdescribes electricity, then, because there is nothing meeting his description, he is not talking about electricity at all.
- **H.** Similarly, this descriptive conception of reference looms large in the somewhat exaggerated incommensurability arguments of Kuhn and Paul Feyerabend. If enough descriptive content changes, the reference will likely change with it. Thus, when descriptions of mass change across theories, the new theory often refers to something new, namely, mass-as-conceived-by-the-theory. For this reason, Einstein cannot offer a better theory of the same mass as Newton's, and this makes progress and accumulation difficult.

II. A new conception of reference emerged (mainly in the 1970s) that makes it easier to talk about unobservable reality and to keep talking about the same things or properties, even across major scientific changes. On this view, reference (for certain kinds of terms) is secured through a historical chain, rather than through a description. It is often called a *causal theory of reference*.

A. Proper names provide the easiest starting point. If you say, "James Buchanan, the 14th president" (he was actually the 15th), you are still referring to Buchanan.
   1. Buchanan's name was attached to him via a kind of baptismal event, not a description. This is a stipulation.
   2. My use of his name is linked to previous uses in a causal chain that terminates in the baptismal event. I intend to refer to the same man as the person from whom I learned the name, and so on, back through the chain to the first link.

B. Similar things can be said of "natural-kind" terms, such as biological species. We would like a theory that allows us to say that people who thought that whales were fish nevertheless referred to whales.
   1. The reference of such terms gets fixed via an archetypal specimen: Whales are creatures like this one.
   2. *Like this one* means having the same deep or essential properties. For chemical elements, it will be their atomic numbers.

C. There is a division of linguistic labor involved in this picture. I do not have to know much about James Buchanan in order to talk about him. Similarly, I do not have to know deep facts about whales in order to succeed in talking about them.

D. This new conception of reference had an unexpected consequence: It helped make metaphysical discourse look more respectable than it had to the positivists.
   1. If Hesperus and Phosphorous are two different names (rather than descriptions) for the planet Venus, then it is necessarily true that Hesperus is Phosphorous, and this is not a necessity that is analytic and knowable *a priori*. Room is made for a notion of metaphysical necessity that does not reduce to conceptual necessity.
   2. This talk of a deep structure shared by all members of natural kinds, such as chemical elements, also

rehabilitates, to a significant extent, the notion of essences, which had long been thought unduly metaphysical. These deep structural properties look scientifically respectable.

E. This approach to reference also makes incommensurability look much less threatening than it had. Insofar as this approach can be made to work, theory change, even across revolutions, can involve competing theories about the same "stuff," rather than just theories about different "stuff."

F. The causal/historical approach does make it easier to talk about unobservable reality in a meaningful way. On the assumption that water has a deep structure responsible for its nature, the historical chain approach allows one to talk meaningfully about that structure.

G. However, we can never encounter specimens of the purported objects of some theoretical terms. We cannot point at an electron and say, "I mean to be talking about everything that is like that thing." Given how messy the notion of causation is and how messy the causal chain would have to be, it would be hard to pick out an electron as what is responsible for the streak in the cloud chamber.

H. The historical chain approach can also make it too easy to refer to unobservable reality. We don't want to count someone as referring to oxygen when using the term *phlogiston*, even though oxygen is what is causally responsible for combustion.

III. A new conception of scientific theories also makes it easier to extend meaning and reference to unobservable reality.

    A. The received view of theories treats them as deductive systems, which get interpreted when some terms are explained experientially. Statements involving theoretical terms generally receive only a partial interpretation.

    B. A newer conception of theories draws on the notion of a model.

        1. A model can be formal. For instance, a wave equation can be used to model waves of sound, or of light, and so on.

- **2.** Models can also be material, in which case they interpret the theory in terms of real or imaginary objects, rather than abstract structures. For example, gas molecules are modeled as small, solid balls.
- **C.** Logical positivism assigns only a modest role to models.
    - **1.** Models can serve a heuristic function. They involve pictures or analogies that are useful for understanding a theory or for using it.
    - **2.** But the model is not part of the theory, and the theory, not the model, is what says what the phenomena in its domain are like.
- **D.** But if the model continues to be useful in enough different contexts, it becomes more than just an aid or a supplement to the real theory. A good enough model virtually becomes the theory. Models loom large in scientific practice.
- **E.** The *semantic conception of theories* identifies a theory with the entire class of its models. A correct theory will have the real world as one of its models. An ecological theory can be interpreted, for example, via patterns of shapes and colors on a computer screen, or via mathematical equations, or via actual patterns of fox and rabbit populations.
    - **1.** The big departure from the received view is that semantic approaches allow theoretical terms to be interpreted directly through models, rather than requiring that interpretation always arise through observation.
    - **2.** The semantic conception thus allows a role for analogical and metaphorical reasoning in science. These types of reasoning can provide literal content to what our theory says about unobservable reality.
- **F.** But how do we restrict the permitted types of modeling and analogical reasoning?
    - **1.** What stops someone from claiming to understand absolute simultaneity on the model of local simultaneity?
    - **2.** With some theories, most notably quantum mechanics, there seems to be powerful reasons to resist taking models too seriously.

Essential Reading:

Putnam, "Explanation and Reference," in Boyd, Gasper, and Trout, *The Philosophy of Science*, pp. 171–185.

Kitcher, "Theories, Theorists and Conceptual Change," in Balashov and Rosenberg, *Philosophy of Science: Contemporary Readings*, pp. 163–189.

Supplementary Reading:

Spector, "Models and Theories," in Brody and Grandy, *Readings in the Philosophy of Science*, pp. 44–57.

Questions to Consider:

1. Do you think that science should strive to be as free of metaphor and analogy as possible? Why or why not?
2. Suppose there were a substance that behaved just like water (for example, we could drink it) but had a quite different molecular structure. Would that substance count as water? Why or why not?

Lecture Twenty-Five—Transcript
New Views of Meaning and Reference

We've now surveyed the classic positivist views of explanation, causation, laws of nature, and scientific reduction, and we've looked at some of the major challenges these views face. At this point, we can begin bringing our two narratives together (not that we ever kept them all that far apart) by integrating issues about meaning and reference more fully and directly into the discussion.

This will allow us, in the next lecture, to begin to develop scientific realism, the major contemporary alternative, along with a kind of Kuhnian historicism or constructivism, to the views of the positivists with which we began.

The difference between meaning and reference figured in our discussion of reduction. Temperature, we learned, doesn't mean the same thing as average molecular kinetic energy, but it is, as it were, the same stuff. The two terms co-refer. Technically speaking, noun-like terms (names and descriptions) refer to things, while adjective-type terms technically don't refer; these are predicates that hold of, or are true of things, rather than referring to them.

So "Albert Einstein" and "discoverer of special relativity" refer to the same person, though they don't mean the same thing, while "cordate" (which means creature with a heart) and "renate" (which means creature with kidneys) hold of all the same things, though they don't mean the same thing. But I'm not going to respect this technical notion about "holding of" versus "referring to" very much. I'll talk rather informally about such matters.

Phrases, as well as words, can refer. As we saw in our discussion of Bertrand Russell's "present King of France" case, a phrase like "the favorite physicist of the logical positivists" has to pick out a unique individual in order to refer (that's why the phrase "the present King of France is bald" is false; it doesn't pick out a unique present King of France). If a description picks out a unique individual (let's say that Einstein is the favorite physicists of the logical positivists) then the description "the favorite physicist of the positivists" refers to the individual whom it correctly describes.

But now suppose that Werner Heisenberg, the great theorist of quantum mechanics, actually turns out to have been the favorite physicist of the logical positivists. In that case, I might think I'm using that phrase to refer to Einstein, but, in fact, I'm referring to Heisenberg. Similarly, if I, in fact, refer to oxygen when I use the phrase "substance responsible for combustion," but if the phlogiston theory had been correct, then I would have been referring to phlogiston rather than oxygen. Our words, rather than our intentions, pick out what it is that we're talking about. Sometimes we're not talking about what we think we're talking about; that's a reasonably common phenomenon.

As we've seen, the positivists treated meaning and reference as relatively unproblematic for observation terms, but as quite problematic for theoretical terms. We also saw them starting to modify that view a little bit by acknowledging at least some of the holistic pressures that were later emphasized by Quine, Kuhn, and Feyerabend.

But the sort-of radical "hard core" positivists' approach does not provide reference for theoretical terms at all; the parts of scientific theories that aren't about experience don't directly refer to the world and don't aspire to truth; they're only partially interpreted. Talk of quarks serves not to describe the world, but to systematize and to predict observation.

There are less-stringent empiricists (and some of the positivists became less-stringent empiricists) who allow theoretical terms to refer and treat them in the standard sort of way. So, a phrase like "smallest unit of a chromosome capable of undergoing mutation" refers to whatever thing—if there is one—that uniquely meets that description. This is also the approach taken by Kuhn and like-minded thinkers, though with an important difference. For Kuhn, reference is surprisingly easy to secure, because for Kuhn, a term refers only to the world as described by a paradigm. So, in Kuhn's view, a term like "phlogiston" does refer to something (standardly, we might think it doesn't refer to anything at all since there is no phlogiston) just as surely as oxygen does. Both refer to paradigm-constructed worlds rather than to some notion that Kuhn thinks is empty of "the world."

Kuhn's view threatens to make successful reference a little bit too easy. Within a certain theory, "the little green men who are out to get me" does refer to something real—something real within my paranoid delusion. I'm not accusing Kuhn of lacking resources for distinguishing a serious scientific paradigm from a paranoid delusion, but even with a serious scientific paradigm, most philosophers find it more natural to say that phlogiston never existed, and so the term "phlogiston" never referred to anything, as reasonable as it might have been to think it once referred. Most philosophers think it adds an unnecessary layer of confusion to talk about referring to something within a paradigm. It seems simpler and logically cleaner to just talk about whether there's any such stuff—but that's not Kuhn's view.

On the other hand, if you stick with the more intuitive notion of successful reference (to the world, not to the world as described by a paradigm), then this descriptive conception of reference makes it very hard (where Kuhn had made it very easy) to secure a legitimate reference. So, if Benjamin Franklin misdescribes electricity while performing some of his experiments, then—since there's nothing at all in the world (as opposed to in the world as described by his paradigm) that actually meets his mistaken description of electricity—it turns out he's not talking about electricity at all. He's not referring to anything because in order to refer to anything, he has to pick it out correctly in his description. So, poor Ben isn't talking about anything at all.

Similarly, this descriptive conception of reference looms large in the, as we've seen, somewhat exaggerated incommensurability arguments put forward by Kuhn and Feyerabend. If enough descriptive content of a theory changes, reference is likely to change with it. So, as Kuhn emphasizes, when descriptions of mass change across theories, the new theory refers to something new—namely, mass as conceived by the new theory rather than a quite different stuff, mass as conceived by the old theory. So, Kuhn argued that Einstein isn't offering a better theory of the same mass that Newton had been theorizing about—but, rather, a theory of a different mass. It's in part for that reason that Kuhn and Feyerabend thought that progress and accumulation were very rare—if ever—instantiated in science.

But those arguments seemed to us at least a bit exaggerated. What we'd like is a view that allows us to refer to something real even when we have some false beliefs about it (and use false descriptions about it)—and also a view that makes reference a matter of the way the world is, rather than the way our theory says the world is. In short, what we need is a conception of reference that is less closely tied to our current theories than the descriptive account presents reference as being.

A new approach to reference emerged mostly in the 1970s that makes it much easier to talk about unobservable reality (to talk seriously about unobservable reality, in fact) and, not accidentally, to keep talking about the same things or properties, even across major changes in scientific theories. On this view, reference (at least for certain kinds of terms) is secured through a historical or causal chain, rather than through a description. This is often called the *causal theory of reference*, and two American philosophers, Saul Kripke and Hilary Putnam, are the main innovators here.

The easiest way to start understanding this theory is to begin with the notion of proper names. Few of us can offer much in the way of true descriptions of James Buchanan. How is it that we can refer to James Buchanan, even though we can't say very much about him other than that he's a former President? In addition, somebody who uses the name along with some false descriptions (so, for instance, somebody who says "James Buchanan, the 14th President of the United States" is referring to James Buchanan, not to Franklin Pierce—who is actually the 14th President of the United States [Buchanan is the 15th]) still refers to him rather than that person who actually fits the description. That's a problem on the descriptive theory, which says that our reference is to whatever fits the description. If I say "James Buchanan, the 14th President," I'm talking about James Buchanan, not about the 14th President.

The idea is that Buchanan's name got attached to him not via a description, but via a kind of baptismal event. The idea is there's a stipulation—his parents or some authority, in some sense, said, "Let's call this person James Buchanan."

My use of that name is linked to previous uses in a causal chain that terminates in this original baptismal stipulation. I intend to refer to the same guy as the person from whom I learned the name (some history teacher in junior high or something) and so on, back through the causal chain. My reference is secured through the reference of other people.

Turning to more scientifically relevant cases, the idea is that similar notions of reference apply to what are sometimes called *natural kind* terms. The main exemplars of natural kind terms are biological species and chemical elements. A natural kind is a grouping with a deep, scientifically relevant structure. It's objects that share a scientifically real property. If nature has any joints, natural kinds carve the joints. So, for instance, gold—which is a chemical element—would be a natural kind; precious metal might be a legitimate term for some purposes (say, economic purposes), but it's not a natural kind—precious metals need not have any deep structure in common.

So, we'd like, with natural kind terms, to have a theory that allows us to say that people who thought whales were fish, nevertheless, were referring to whales (especially if they're pointing at a whale when they say, "Wow, what a big fish"). The fish description does not fit the whale, but we'd like them to be saying false things about the whale, rather than referring to nonexistent, great big fish.

A species term isn't quite like a proper name in that it designates a class or a kind rather than an individual. So, the reference of the term gets fixed via a kind of archetypal specimen. You point at a whale, and you don't have to literally say it, but what's going on with the baptismal event is something like "I mean creatures like this one." The "like this one" means having the same deep or essential properties.

For chemical elements, it will be their atomic number. What it is to point to a sample of gold and say "I mean to be talking about that kind of stuff" is stuff that shares the atomic number, not stuff that shares superficial properties like being yellowish.

As we'll see later in the course, it's not clear that species really fit this bill of natural kinds all that well, though philosophers constantly use species of examples of natural kind terms.

Another interesting aspect of this view is that there is a division of linguistic labor involved in how reference is secured. I don't have to know very much about James Buchanan in order to talk about him; I just have to be connected to the causal chain in the right sort of way.

Similarly, I don't have to know the deep facts about what makes a whale a whale in order to succeed in talking about them. My community's experts (certain biologists, in this case) help fix the reference of my terms. Our best science decides what the deep structure is that constitutes the kind. So science tells me, in part, what I mean when I point at whales and say, "I mean those kinds of creatures."

As on the descriptive view, there's room for people to not know what they're talking about, which is a good thing, since we'd like our discourse to outrun our knowledge. I don't need to know what I'm talking about when I talk about whales; somebody in my linguistic community can say, "Here's what makes a whale a whale." I don't need to know it in order to be talking about whales.

This new conception of reference had some unexpected consequences. It helped make metaphysical discourse look more respectable than it had to the positivists. This is part of the beginning of the end of positivism.

If "Hesperus" and "Phosphorus" are two different names (rather than descriptions; descriptions raise different issues) for the planet Venus, then it's necessarily true that Hesperus is Phosphorus, since it seems necessarily true that anything is itself. But this is not a necessity that is conceptual. It's not analytic; it's not knowable *a priori*. It was a discovery that the morning star was the evening star—that Hesperus is Phosphorus.

Room is made for a notion of metaphysical necessity that doesn't reduce to conceptual necessity. Most philosophers agreed that identity seemed like a solid-enough basis for a necessary truth, but it's not a necessity grounded in the meanings of terms, but instead, in their references. So, there's a kind of necessity that's not grounded in meaning as the positivists had claimed that all necessity would have to be; there's a kind of necessity out in the world. This does not, by any means, establish the kinds of physical necessity that we've appealed to in the past in this course. For instance, the physical necessity by which some philosophers think being made of copper necessitates conducting electricity. But it does open up the idea of necessity out in the world rather than in our language, and that's a real problem for positivists.

This talk of a deep structure shared by all members of a natural kind like a chemical element also, to a significant extent, rehabilitates the old notion of essences, which had long been thought the very model of metaphysical excess. A kind of structural essence that figures in scientific explanations looks scientifically respectable. Explaining gold in terms of its atomic number looks like the essence of gold, what makes gold gold. So, positivism—again—had to start retreating. The idea is experience doesn't mark some properties as essential to a substance, leaving others as accidental (we don't see what makes gold gold), but the role of some properties in science suggests a difference between deep properties and superficial ones, between essential and accidental properties. Yellowness is an accidental property of gold; the atomic number is an essential or deep property of gold. This is almost a reversion to Aristotle.

This approach to reference also makes the incommensurability arguments of Kuhn and Feyerabend look much less threatening than it had. Insofar as this approach can be made to work, theory change—even across deep scientific revolutions—can be said to involve competing theories about the same "stuff" rather than just theories about new and different "stuff." It's a pretty big change if we go from thinking of light as a particle to thinking of light as a wave. If we're adopting the descriptive conception of reference, it looks like we're talking about two different things. But on the causal or historical approach, it's pretty straightforward to insist that both are theories about light. Theorists of both camps intend to refer to the stuff that previous theorists of light were talking about. So, they are competing theories about the same stuff, so one can be a better theory of this stuff than the other.

This historical chain approach also offers a set of alternatives to those who had felt hampered by the empiricist strictures on meaning—including those, for instance, who thought that empiricist notions of meaning (things that were common to Kuhn and the positivists) made incommensurability a bigger problem than it needed to be. But the causal/historical approach is not itself without problems (of course, by now, you know no philosophical view is without problems). When we try to apply it to theoretical terms, we're going to run into some difficulties.

This approach does make it relatively unproblematic to refer to some aspects of unobservable reality. On the assumption that water has a deep structure responsible for its nature, the historical chain approach allows me to talk meaningfully about that structure. I point at some water and say, "I'm talking about stuff like that." To a first approximation, I can refer to anything that is directly causally responsible for anything I can observe (I can get behind observation, at least one level, through something like causation).

But we can never encounter specimens of some of the purported objects of theoretical terms. We can't point at an electron and say, "I mean to be talking about stuff like that," because the way an electron manifests itself in experience is incredibly messy. There are all sorts of mediating links in the causal chain; to point at a cloud chamber and say, "I mean whatever's responsible for that," there are too many ways the apparatus could be misfiring. You could end up talking about some experimentally induced artifact rather than the electron.

The historical chain approach can also make it too easy to refer to unobservable reality (this is a kind of positivistic worry about the approach). "Phlogiston" might count as referring to oxygen if we're not careful, because if we mean to be referring to what's casually responsible for combustion, well, that's oxygen. But we don't want to say "by "phlogiston," I mean "oxygen." They're supposed to be competing theories of the same phenomenon.

Similarly, it might have been possible for the ancient Greeks to refer to electricity. If they got a shock of static electricity and said, "I wonder what that stuff is," on the causal chain approach, they might count as referring to electricity. But they knew nothing of electricity. It's not clear that we want to let reference float entirely free of connections to descriptions. Arguably, you need to know something about the kind of stuff electricity is before you count as talking about it. Maybe you don't have to know much, but you need to know something about it.

Because of the problems just noted, descriptive accounts of reference have been making a comeback in recent years. Aspects of the descriptive and historical chain approaches to reference can, in fact, be combined. So, for instance, you might think that you need enough of a correct description to pick out a unique object that is causally connected to what's observed (that's to combine some of the descriptive requirements with some of the causal historical requirements).

But it's not as easy as it sounds. If too much descriptive content is built in, then the term might not refer to anything, because nothing will satisfy the description. We don't want Newton's term "mass" to refer to nothing at all, even though there's nothing in the world that answers to all of Newton's descriptions about mass.

But if too little descriptive content is built in, then there will be no unique satisfier of the description. We want a term like "cause of combustion" to refer to oxygen, but not to phlogiston.

So, the causal/historical approach provides some promising resources for enriching our notion of semantic access to the world. We can talk about unobservable reality in a way that we couldn't under the positivist conception of meaning. But the reintroduction of descriptive requirements (as had figured in positivism) means that we have not entirely escaped the problems with which we began. There's still some room for incommensurability worries, and we still have rather modest resources for talking about unobservable reality. That's at the level primarily of terms.

A similar broadening of semantic resources occurs through a new conception of scientific theories.

As we saw way back in Lecture Seven, the received view of theories treats them as deductive systems. The statements of a deductive system are themselves uninterpreted; they are purely syntactic (they exhibit logical relationships, but nothing more).

The deductive system then gets interpreted when observational terms get explained experientially. All interpretation, for the positivists, comes through observation. The logical structure of the theory then lets meaning "flow" around the system. Many statements of the theory only receive a partial interpretation in observational terms (so they're not straightforwardly true or false). Statements involving theoretical terms are construed as something like inference tickets, not as descriptions of the world.

This newer conception of theories draws on the notion of a model. Models represent things by exhibiting structural similarities to the things they model. So, a modeling relationship is a type of analogy; it displays certain similarities between phenomena of different types.

So, for instance, a model can be formal. A wave equation can be used to model waves of sound, waves of light, and waves of water. Models can also be material, in which case they are interpreted in terms of objects—real objects or imaginary objects—rather than abstract structures, as in a mathematical model. In the classic cases, gas molecules are modeled as small balls of plastic or something like that, and atoms are modeled on the solar system, with electrons orbiting the nucleus.

Models have been around in science for a long time. The philosophical questions concern their status and role. The positivists didn't ignore models, but they assigned only a fairly modest role to them.

Models for positivism serve a kind of heuristic function. They are pictures or analogies that are useful for understanding a theory and may be helpful for using it in, say, experimental design. But they belong to what the positivists called the context of discovery, not the context of justification. It's fine to think of gas molecules along the lines of little wooden balls if that will help you picture what's going on. It can help you figure out how to test hypotheses and how to draw some connections. But the model is not part of the theory, and the theory is the only thing that says what gas molecules are like. The model is an add-on that might help you do some things with the theory, but it's no part of what the theory says.

One impetus for an increased role of models in thinking about science was their pervasiveness in actual scientific practice. It's much more common for scientists to present their theories through models than through deductive systems. In keeping with the post-Kuhnian emphasis on actually representing scientific practice accurately, models started to receive more attention than they had in the heyday of positivism.

So, for critics of the received view (semantic theorists, model theorists of scientific theories), a model might start out as just a way of picturing or thinking about an experimental situation. But if the model continues to be useful enough in enough different contexts, it becomes more than just an aid or a supplement to the real theory.

To take an example from Marshall Spector, a philosopher of science, nobody would mistake acoustical flow for electrical flow. They're clearly different things. But the analogy between them starts to seem more than just a kind of mathematical similarity. The notion of impedance or capacitance seems to mean the same thing in both theories, and the notion—as it figures in one theory—can be used to illuminate the notion as it figures in another. The received view of theories has a hard time saying that the notion of flow or capacitance means the same thing in two different theories. So, the model starts to say things that the theories, in the received view, have a hard time saying.

If the model is good enough, it more or less becomes the theory. That won't happen in the case we were just talking about. We'd never think our theory of electrical flow is our theory of acoustical flow. But at some point, we stop just picturing gas molecules as little balls—that's how the theory actually functions in scientific practice, at least within a certain range. That's what scientists think gas molecules are like. They don't think the gas molecules have color like a plastic ball does, but they think they are little balls undergoing inelastic collisions. The model is not an aide to the theory—it's closer to being the theory.

So, the *semantic conception of theories* (the new approach to scientific theories) identifies a theory with the entire class of its models. (This is a part of philosophy of science that gets very technical very quickly. We're going to skate over the technicalities.) The theory includes all the ways in which it can be modeled. A correct theory will have the real world as one of its models. A correct interpretation of the theory in terms of the real world means that the theory is true.

An ecological theory can be interpreted, for instance, through patterns of shapes and colors on a computer screen, through mathematical equations, or through actual patterns of fox and rabbit populations. Each of these is a legitimate way of giving literal content to the theory's laws or claims.

This represents a major departure from the received view of theories because this semantic approach to scientific theories allows theoretical terms—terms that can't get "cashed out" directly through observation—to acquire full-fledged literal meaning through the notion of a model, rather than through observation. The notion of flow in acoustics gets genuine meaning from the notion of flow in electrical circuit theory, for instance, or vice versa.

The semantic conception of theories thus allows a role for analogical and metaphorical reasoning to do genuine work in science. These types of reasoning can provide literal content to what our theory says about unobservable reality. This had been forbidden on the received view; all literal content had to ultimately get "cashed out" in observational terms.

So, an approach like this will seem promising to someone who felt hemmed in by the empiricist strictures on meaning. But, as always, there's a price to be paid. We are helping ourselves to enriched resources for referring to unobservable reality. In doing so, we run a risk of falling into the mistakes the positivists had tried to warn us about: How are we to restrict the permitted types of analogical reasoning? We don't want to be able to claim to model absolute simultaneity on local simultaneity. This would be to have unlearned the lesson that Einstein was trying to teach us: We shouldn't help ourselves to some understanding of absolute simultaneity because it can't be "cashed out" experientially. The motivation for the empiricist project was that we confuse ourselves if we think meaning can be stretched too far beyond experience. But the semantic conception thinks we can stretch it farther beyond experience than the received view allowed.

With some theories, most notably quantum mechanics, there do seem to be powerful reasons to resist taking models too seriously. This is a theory such that it's dangerous to interpret it much beyond its mathematical content. Even heuristic models that are only pictures to help us understand the theory threaten to introduce elements that are incompatible with the mathematics of the theory.

There are some models of quantum mechanics that are suggestive and careful, but if you're not extremely cautious as you try to picture or model what's going on, you're going to find yourself tempted to posit what physicists call *hidden variables*, something behind the scenes to make the theory more deterministic and so more intuitively comprehensible than the mathematics, as classically interpreted, will let you interpret it.

It's hard to model quantum mechanics on anything less weird than quantum mechanics itself. If we do model quantum mechanics on something less weird, we face the danger of giving a particle a definite position and momentum, which it seems intuitively like it would have to have, but which the classic interpretation of quantum mechanics says it can't have. If we model quantum phenomena, we're tempted to try to find a model according to which the electron passes through one slit or another, because it seems to us that it's got to behave that way. But the mathematics of the theory, as classically interpreted, don't permit that interpretation. So, there's a danger in using these enriched semantic resources to bringing in causal intuitions that the theory says are unsupportable. So, there are advantages and disadvantages to these enriched semantic resources.

These new conceptions of meaning, reference, and scientific theories hopefully will extend our semantic reach beyond what the positivists had permitted. In doing so, we'll start to see the difference this might make to the aims of science. Can science aim to describe unobservable reality correctly? Once we start taking that claim seriously, we're going to witness (more or less) the end of logical positivism.

Lecture Twenty-Six
Scientific Realism

Scope:

The semantic developments sketched in the previous lecture make room for the doctrine of *scientific realism*, which requires that science "talk about" unobservable reality in much the same way that it talks about observable reality. In this lecture, we examine the varieties and ambitions of scientific realism, contrast it with empiricism and constructivism, and confront two major challenges to realist interpretations of science.

Outline

I. A number of considerations convinced many philosophers that there is no interesting distinction to be drawn between observational and theoretical language. Without such a distinction, logical positivism is more or less dead. The epistemology of empiricism can live on, but it will have to take a different form (as we will see).

   A. The new conceptions of meaning and reference that we canvassed in the last lecture suggested that our semantic reach can extend farther beyond observation than the positivists had thought.

   B. A relatively modest descendant of a point made by Kuhn and Feyerabend also contributed to the new skepticism about the observational/theoretical distinction. They insisted that theories shape what we see and how we describe what we see.

   1. Most philosophers were not enormously impressed by the argument that our theories "infect" our observations. By and large, philosophers accepted only modest versions of this claim.

   2. But they did become convinced that our theories "infect" our observational *language*. We use theoretical terms (such as *radio*) to talk about observable things. Such talk is fully, not partially, meaningful. The majority of philosophers gave up on the idea that anything worth

calling science could be done in a language that was sanitized of reference to unobservable reality.

C. Conversely, we can use observation terms to describe unobservable objects (as when we picture gas molecules as little billiard balls).

D. Thus, the distinction between observable and theoretical language does not line up with the distinction between observable and unobservable objects.

II. Statements about unobservable reality, then, can be true or false in the same way that statements about observable reality can. This makes room for *scientific realism*, a view that requires that science aim at accurately depicting unobservable as well as observable reality. What else is involved in scientific realism?

A. Metaphysical modesty is a requirement: The way the world is does not depend on what we think about it.

B. Epistemic presumptuousness is also a requirement: We can come to know the world more or less as it is.

C. Although each of these theses holds considerable appeal, they tend to work against each other. The more independent the world is of us and our thought, the more pessimistic it seems we should be about our prospects for knowing it.

III. We have seen two anti-realist positions that reject metaphysical modesty, and these can be compared with two realist positions that accept different versions of metaphysical modesty.

A. The logical positivists reject questions about the way the world is. They consider such questions invitations to metaphysics.

B. For Kuhnian and other constructivists, the way the world is *does* depend on what we think about it.

C. For "hard" realists, *the way the world is* means that some distinctions, similarities, and kinds are, as it were, "out there." The world determines that gold is a real kind, all the instances of which share important properties, while *jade* names an unreal kind, two different kinds of things (jadeite and nephrite) that go by one name.

D. For "soft" realists, *the way the world is* means only that, given certain interests and aptitudes, it makes good sense to categorize things in one way rather than another (for example, to think of gold as one kind of thing, but jade as two). Our best theories take our interests into account, but they are still responsible to a mind-independent world.

E. Hard realists think that the job of science is to find out the way the world truly is, and this goal has nothing to do with contingent human limitations. Soft realists think that the aim of science is to organize a mind-independent world in one of the ways that makes most sense to us. Soft realists generally permit the idea that incompatible theories could be equally good, while this is much harder to grant according to hard realism.

F. Hard realism runs the danger of being too restrictive, while soft realism can easily become too permissive. As we will see in later lectures, it's not clear that the world has many kinds that live up to hard realist standards. On the other hand, not every classification scheme that's good for certain purposes thereby gets to claim that the classification is correct.

IV. Turning from metaphysical issues of modesty to epistemological issues of presumptuousness, we can review some previously examined positions and compare them to a couple of versions of scientific realism.

A. Logical positivists think that we cannot get evidence that bears on the truth of statements about unobservable reality. Therefore, we should not presume to have knowledge that so thoroughly outruns the evidence.

B. For Karl Popper, it is possible that we could come to know the world as it is, but because there is no usable notion of confirmation, we'll never be in a position to claim such knowledge about anything.

C. For Kuhn and other constructivists, knowledge of the way the world is would require stepping out of our intellectual and perceptual skins. Even if the project made metaphysical and semantic sense, it would be excessively epistemically presumptuous.

D. For "optimistic" realists, our best scientific theories provide knowledge of the way the world is (including unobservable reality). This is the most epistemologically presumptuous view out there, but it's not a crazy or uncommon one. However, this view sets things up so that if major scientific theories are false, then scientific realism is false, and that seems undesirable.

- E. For "modest" realists, it is reasonable to hope that science can, and sometimes does, provide knowledge of the way the world is. Such thinkers count as realists because they think science has a reasonable chance of getting the world right, but they need not think that it has done so.

V. The most important debates among realists and between realists and their opponents have concerned epistemic issues: How confident should we be that science does, or at least can, provide us with knowledge of unobservable reality?
- A. The *underdetermination of theory by data* made its first appearance in W. V. Quine's work.
    1. It is often the case that all the currently available evidence fails to decide between two competing theories. But this needn't trouble the realist much so long as science has some decent prospect of determining which theory is true.
    2. Stronger versions of underdetermination claim that all possible evidence underdetermines theory choice. This is awkward for the realist, who needs to claim that (at most) one of the theories is true.
- B. A couple of replies are available to the realist.
    1. One is to deny that we can always find genuine theories that compete with a given theory. For example, I would not be proposing a new theory if I switch the terms *positive* and *negative* so that electrons have a positive charge and protons have a negative charge. This is the same theory in a verbally incompatible form.
    2. Realists can also appeal to principles governing the way to run a web of belief and claim that of two theories that fit the data equally well, one might, nevertheless, receive more evidential support than the other.
- C. The other major obstacle to realism is an important historical argument called the *pessimistic induction*.

1. We can find cases from the history of science of theories that did as well or better than current theories by the best evidential standards of the day. Because we now know those theories to be false, we should not think our best theories likely to be true.
   2. This objection follows Kuhn in thinking that the history of science is our best guide to how science should be done. But it tries to demonstrate that history shows that realism is unwarranted, because the best standards of actual science permit false theories to thrive.
D. The realist has room to maneuver here, as well.
   1. If some version of the traditional approach to scientific reduction can be defended, then one can claim that superseded theories are preserved by being reduced into superseding theories.
   2. Realism might need to narrow its ambitions and claim only that parts of our best theories are likely to be true or that only some of our best theories are likely to be true. Not all aspects of our theories are equally accessible to us or equally well tested.
   3. Realism could be defended concerning the mathematical structures involved in our best theories, rather than the entities posited by them. Nicolas Carnot worked out many of the basic ideas of thermodynamics, despite the fact that he mistakenly thought of heat as a kind of fluid.

Essential Reading:

Nagel, "The Cognitive Status of Theories," in Balashov and Rosenberg, *Philosophy of Science: Contemporary Readings*, pp. 197–210.

Laudan, "A Confutation of Convergent Realism," in Balashov and Rosenberg, *Philosophy of Science: Contemporary Readings*, pp. 211–233 (also in Boyd, Gasper, and Trout, *The Philosophy of Science*, pp. 223–245, and in Curd and Cover, *Philosophy of Science: The Central Issues*, pp. 1114–1135).

Supplementary Reading:

Psillos, "The Present State of the Scientific Realism Debate," in Clark and Hawley, *Philosophy of Science Today*.

Questions to Consider:
1. How sympathetic are you to the idea that science does (or at least can) "carve nature at its joints"? What considerations could help you decide between a hard realism like this and a soft realism or an anti-realism?
2. How independent of thought does the notion of "the world" or "the truth" seem to you? Surely my thinking something doesn't make it so. But what about the idea that any statement that would be agreed upon "at the end of inquiry" would have to be true? Is this conception of truth too metaphysically immodest? Why or why not?

Lecture Twenty-Six—Transcript
Scientific Realism

The sorts of considerations we examined last time promised to extend our semantic reach (the meanings of our terms and theories) beyond observation in a way that the logical positivists and their successors, like Thomas Kuhn, had not envisaged. These ideas, first, that reference can be secured even when we can't use observation to assign content to a term or statement; and second, the idea that models and analogies can be used to provide literal meaning for terms and statements at a genuine remove from observation—these two ideas together formed part of an attack on the importance and even the intelligibility of the distinction between observational and theoretical language.

Without such a distinction, logical positivism is more or less dead. The epistemology of empiricism can live on, but it will have to take a deeply different form if the distinction between observational and theoretical language is surrendered.

A relatively modest version of a point made by Kuhn and Feyerabend also contributed to the new sort-of skepticism about the interest and importance of the observational/theoretical distinction. We saw Kuhn insist that theories shape what we see and how we describe what we see. We saw Feyerabend take this point and run with it rather farther than Kuhn had, almost minimizing the role of observation in science.

These theses were more extreme than any that caught on with philosophers, though they did do well with constructivists and postmodernists outside the discipline of philosophy. Though most philosophers accepted some version of the theory-ladenness of observation, they did not think that it was so pervasive that we couldn't have theoretical language figuring in observation statements that nevertheless can exert an epistemic function independently of the theoretical language in which they're formulated. So, a pre-Cambrian rabbit would make mischief for Darwinian evolution, even though such an observation is inescapably couched in the language of the theory that it would ultimately make mischief for.

Philosophers did, however, open up to the idea that whether or not theories—in any important way—"infect" our observations, they do pervasively infect our observation *language*. The vast majority of philosophers eventually gave up on the idea that anything worth calling science could actually be done in a language that was sanitized of reference to unobservable reality, as the positivists had tried to sanitize science. It had long been understood that science isn't actually done in a language sanitized of reference to observation—scientists talk about unobservable reality all the time. The issue concerned the logical status, the rational reconstruction of science, not the actual practice.

But for that very reason, the increasing emphasis on actual scientific practice threatened the rational reconstruction—according to which theoretical terms have a kind of second-class semantic status—with irrelevance. Then problems started to emerge with the reconstruction itself. Whatever its importance, philosophers started to worry about the intelligibility of the rational reconstruction.

For one thing, we use theoretical terms to talk about observable things. Radios are observable objects, but describing an object as a radio involves going beyond the observation language, since you're calling it a receiver of radio waves, and radio waves are unobservable. So, we can use theoretical *language* to describe observable *things*.

Conversely, we can use observation terms to describe or refer to unobservable objects, as we do when we think of gas molecules as little billiard balls, or when we talk about a speck of dust that's too tiny to see. All of that language is observation language, but it refers to an unobservable object.

So, the distinction between observable and theoretical discourse does not line up with the distinction between observable and unobservable objects.

So, with this distinction seeming none too clear, and in light of the semantic developments we looked at last time, theoretical discourse started to be seen as fully—not partially—meaningful. We saw that the positivists did not take apparent reference to unobservable reality with full seriousness. Saying that the top quark has a certain mass is rather like saying that the average American family has a certain level of debt. Just as you haven't committed yourself to the actual existence of an average family in the second case, you haven't committed yourself to the actual existence of a top quark in the first.

All you've done is found a language that's convenient for describing or systematizing observation. For the positivists, apparent reference to unobservables has mostly a kind of bookkeeping function. But that's the view that starts to get very much on the defensive starting, say, in the 1970s or so.

The pervasive role of theoretical discourse in actual scientific practice, even in what working scientists would think of as observation reports (a working scientist would count a statement like "the pH of the solution was found to be N" as an observation report, even though that's very much clothed in theoretical language), this pervasiveness of theoretical language in scientific practice made it increasingly difficult to think of theoretical discourse as somehow semantically inferior to observational discourse.

So, most philosophers eventually gave up on the idea that any interesting distinction between the parts of our language that get meaning directly from observation and the parts of our language that don't, can be sustained. The way we talk, both in science and in ordinary life, takes theoretical pictures of the world quite seriously, and if we tried to drop that part of our talk, we would lose too much of what we'd like to be able to say.

At this point, the received view of theories starts to get called the "once received view of theories." The epistemological virtues that had motivated positivism were starting to seem to carry with them too great a cost, depriving us of too much semantic reach, impairing our ability to do too much of what seemed like good science. That doesn't mean, of course, that reasonably close descendents of the received view of theories haven't survived; they are still very much a powerful presence in philosophy of science—in part because it's difficult to come up with workable alternatives.

But let's say that we've tentatively agreed, for now, to a semantic thesis that would have seemed heretical to the logical positivists. Statements about unobservable reality can be true or false in the same way that statements about observable reality can. It seems fairly intuitive, but not against the background of positivism. Such a semantic doctrine makes room for *scientific realism*, a view that carries with it the idea that science aims at accurately depicting unobservable as well as observable reality. Scientific theories aim to get the whole world right.

First, let's look at one of the crucial terms in this formulation. Though I will continue to talk about the statements of a scientific theory being true or false, that tends to fit well with the received view of theories, according to which a theory is a set of statements. Construing theories as models, the semantic conception of theories, makes a theory more a matter of an analogy. My use of "depiction" in my formulation of scientific realism—that scientific theories aim to depict the world accurately—tries to make room for the notion that we can get the world right in ways other than by saying true things about it. For some philosophical purposes, this distinction matters quite a lot, but I'm mostly going to soft-pedal it. Theories can get unobservable reality right, whether that's by accurately picturing it or by saying true things about it.

This formulation is probably not strong enough, however, since if you thought that science merely aimed at getting theories that accurately depicted the world, but could never attain this aim, you wouldn't count as defending something worth calling scientific realism. So, we're going to need more than the semantic claim that science can describe unobservable reality as well as observable reality correctly in order to develop something worth calling scientific realism.

The first additional requirement we're going to need is one of *metaphysical modesty*: The world, including the unobservable parts of the world, does not depend on what we think about it.

The next requirement we're going to need is one of *epistemic presumptuousness*, or *immodesty*: We can come to know the world, including the unobservable world, more or less as it is.

While each of these statements is intuitively at least somewhat appealing, they tend to work against one another. The more independent the world is of us and our thought, the more pessimistic it seems we should be about our prospects for knowing it. According to metaphysical modesty, much of the world is unobservable by us, but has a nature that doesn't—in any way—depend on us. But epistemic presumptuousness insists that, despite our lack of direct access to this world, at least to big chunks of it, science can manage to form true theories about the world (or at least correct theories about the world), about unobservable as well as about observable reality.

We've seen in this course two anti-realist positions that reject this claim of "metaphysical modesty."

For the logical positivists, asking questions about the way the whole world is constitutes an invitation to metaphysics that should be rejected. Just as a question like Bertrand Russell's famous five-minute hypothesis (which asks us to consider the hypothesis that the universe was created by God five minutes ago, complete with false memories and misleading fossils) that nothing in the universe is actually more than five minutes old, though everything seems like it is, is a meaningless metaphysical hypothesis by the standards of positivism because it makes no difference to expected experience.

Similarly, any questions about unobservable reality should be rejected, or at least translated into questions about observable reality. So, the positivists think we should reject the idea that there's a way the world is (when the "world" there is a notion of unobservable parts of the world). For semantic reasons, we should consider it metaphysical, not scientific, to raise questions about the way unobservable reality is. So, they reject metaphysical modesty mostly for semantic reasons.

Thomas Kuhn and some of his constructivist followers reject metaphysical modesty because they think the way the world is, and the only sense we can make of that phrase, depends on what we think about it. Within a paradigm, questions about the deep nature of reality do make sense, but they make sense of the world according to the paradigm, and that's as deep or genuine a notion of reality as we can make sense of.

So, Kuhn is in agreement with the positivists that we can't make much sense of a notion like "the way the world is" in any extraparadigm sense—when the paradigm changes, for Kuhn, the world itself changes. This seems metaphysically immodest to scientific realists because they think it gives us too much power to determine the way the world is. Our paradigms dictate the way the world is, rather than reflect it.

Similarly, as we'll see a little bit later, realists think that the positivists place too much weight on our sensory capacities, and that seems immodest to them. It's doing too much metaphysical work to talk about what's observable by little creatures like us. The world doesn't respect our epistemic faculties in the way that the positivists thought they would.

So, we can now work in two realist positions, each of which contrasts with these anti-realist positions on the issue of metaphysical modesty—but this contrast works out in importantly different ways.

For what we might call "hard" realists, "the way the world is" means that some distinctions, some similarities, and some kinds are—in a very important and strong sense—"out there." The world tells us (at least if we're doing our science right) that gold is one kind of stuff; it's the world's own distinction between gold and other kinds of stuff, while jade is two kinds of stuff that go by one name. The idea is, all samples of pure gold share the same real deep structure, while samples of jade are samples really of two quite different minerals (I believe they're called jadeite and nephrite), and each of these structures counts equally well as jade. So, jade *is not* a real kind, while gold *is* a real kind.

"Soft" realists, in contrast, rest less weight on the notion of a real kind. So, for a soft realist to say that science tell us "the way the world is" means only that—given certain interests and aptitudes (for instance, given the interests and characteristics of creatures like us)—it makes good sense to categorize things in one way rather than another. It's because of what we care to keep track of that we decide to think of gold as one kind of thing and jade as two. Our best theories take our interests into account, but they are still responsible to a mind-independent world. There is a difference "out there" between gold and jade, but how important that difference is, how real it is, how worth marking it is, is itself responsive to our interests.

So, hard realists think that the job of science is to find out the way the world truly is, and that this goal has nothing to do with contingent human limitations. The goal arises from the nature of inquiry, not from the nature of humanity.

Soft realists think that the aim of science is something more like organizing a mind-independent world in one of the ways that makes the most sense to creatures like us. So, soft realists are generally sympathetic to the idea that incompatible theories could be equally good, while hard realists tend to think that there is a single way the world is, and that science should try to get that "way the world is" right.

There are more sophisticated combinations and versions of these views that can be developed in order to sort-of address the weaknesses that each view at least appears to face. So, hard realism's major problem is that it seems rather restrictive. It's not clear that nature has many joints that live up to the standards of hard realism. The periodic table is a great example, but many other distinctions seem to reflect our need to keep track of things, to make sense of things.

Soft realism, on the other hand, tends to err on the side of being too permissive. Should we allow the category of reptiles to count as a real kind, even though there is apparently no species that is ancestral to all reptiles without also being ancestral to birds? Can we legitimately keep track of that kind, even though it doesn't answer to ordinary standards of biological classification? If we can, then what about the category of cedar trees? Cedar trees are apparently unified by a human interest in a certain kind of wood. The trees that we call cedar trees are not biologically closely related at all. It looks like it might permit too many distinctions or classifications to count as thoroughly real.

Having examined these metaphysical issues and the various flavors of modesty that might be adopted here, we turn to epistemological issues and the various flavors and levels of ambition and confidence that might be adopted on that side of the issue about realism.

As we did with the metaphysical issues, we can now add a couple of new positions to our conceptual geography concerning epistemic presumptuousness. Anti-realists tend not to adopt epistemic presumptuousness (realists do), though there are interesting complicated combinations that we won't be able to cover; I'm just staking out the sort-of clear extremes.

For the positivists, knowledge can extend as far as observation and inductive logic will take us. We'll see in the next lecture what this thesis about knowledge and evidence looks like once it gets uncoupled from the positivists' theses about meaning. So, there are the epistemological theses of the positivists, and there are semantic theses. Realism jettisons the semantic theses—that it makes no sense to talk about unobservable reality. For now, let's just emphasize the point that one can't, for the positivists, get evidence that bears on the truth of statements about unobservable reality; all evidence is observational evidence. So, the positivists reject epistemic presumptuousness: We shouldn't presume to have knowledge that so thoroughly outruns our evidence.

Karl Popper is an interesting case here. For Popper, it is possible that we could come to know the world as it really is, but since there is no usable notion of confirmation, we'll never be in a position to claim knowledge about the way the world is. So, Popper thinks it's not excessively presumptuous to claim that we *can* have knowledge of the way the world is (this distinguishes him from the positivists and from Kuhn), but he does think it's excessively presumptuous ever to actually claim such knowledge. That tends to distinguish him from classic realists.

For Kuhn and other constructivists who also reject epistemic presumptuousness, knowledge of the way the world really is would require stepping out of our intellectual and perceptual skins. So, even if the project made metaphysical and semantic sense, it would be epistemically presumptuous to think that we can shed our biases and the historical ways in which our theories are shaped by preceding theories. So, both the positivists and Kuhn reject epistemic presumptuousness, which is central to scientific realism.

For "optimistic" scientific realists, our best scientific theories (mature, well-tested theories) actually provide knowledge of the way the world is. Our current best theories are at least approximately true, and almost all of the things these theories refer to (quarks, genes, that sort of stuff) actually exist.

This is to adopt a view that's reasonable epistemically presumptuous, but it's not crazy or uncommon to think that our best scientific theories get the world right. The idea is that science not only *can* provide knowledge of the world, it *has* done so, and we can pretty much tell in which cases it has done so. Belief in the truth of our best theories is a major motivation for scientists and philosophers towards realism.

On the other hand, if we build this level of epistemic presumptuousness into our notion of scientific realism, then if our current best theories are false, scientific realism is false—and that seems like a mistake. We'd like to allow for the view that science can get the world right and can reasonably aim to get the world right, without committing ourselves to the view that it has gotten the world right.

So, "modest" realists will say something like that it's currently reasonable to hope that science can, and sometimes does, provide knowledge of the way the world is. We need some reason to think that science has at least a decent shot (this distinguishes such a view from Popper) of succeeding in order for this aim to be reasonable. It's hard for Popper to explain why we should try to get the world right, given that we can never get any evidence that we're doing so.

Within philosophy of science, the most important debates among realists, and between realists and their opponents, have concerned the epistemic issues: How confident should we be that science does, or at least can, provide us with knowledge of unobservable reality?

There are two major styles of arguments that present impediments to realism, one of which we've already encountered.

The first argument is the *underdetermination of theory by data*, which made its first appearance for us in Quine's work. When we first encountered Quine, however, we hadn't yet looked at notions of confirmation and evidence; it was just a matter of a semantic thesis.

As we saw back then, it's often the case that all of the currently available evidence fails to decide between two competing theories. Scientists then try to design experiments that will provide data that bear favorably on one theory and unfavorably on the other. Quine and Kuhn have emphasized how complicated this process is, given the diversity of scientific values and the room to modify various auxiliary hypotheses within one's web of belief. So, the new evidence doesn't decisively favor one theory or the other.

The work of Quine and Kuhn thus suggests that no matter what data we've obtained, there will always be a theory that fits the data equally well. That's awkward for scientific realists because a scientific realist wants to claim that, at least under certain circumstances, one of the good theories is likely to be true. But if the data are equally compatible with each theory, how on earth could we claim to be in a position to decide which of these theories is the true one?

A couple of replies are available to the realist. One is to deny that one can always find a genuine theory that competes with a given theory. As a matter of logic, we can always generate nominally incompatible theories that fit the data. But in many cases, that looks like a logical trick rather than a genuine scientific theory.

The example I'll use is a little bit excessively simple, but it gets the idea across. Let's say I am demanding a Nobel Prize for my brilliant new theory in physics. What I do is I switch, in current physics, every use of the term "negative charge" for "positive charge," and vice versa. So, my theory is different from the current theories, but it's every bit as predictively accurate. We know it's different because I say electrons have positive charge, where the current theory says they have negative charge. So, it's a new theory, and it's very well tested. So, where is my Nobel Prize? It's not really a new theory; it's the same theory in a verbally incompatible form.

So, it's an open question to what extent we'll always be able to generate serious theories that fit the data as well as any given theory.

The other move open to the realist is to appeal to principles governing the way to run a web of belief and the claim that, of two theories that fit the data equally well, one might nevertheless get more support from the very same data—that there's more to being supported by data than merely fitting the data. This is a broadly Popperian notion. One of the theories might have been more thoroughly tested than the other; one might be part of a more progressive research program in Lakatos's sense. Whatever virtue the realist wants to invoke here had better be a mark of truth because the realist wants to say that the better theory has a reasonable shot of being true, while empiricists are going to say the only marks of truth are observational data, and unless you can make that data favor one theory over the other, you're not in a position to claim that one of these theories is likelier to be true than the other one is.

The other major obstacle to scientific realism is an important historical argument (so it's in the broadly Kuhnian tradition) called the *pessimistic induction*. It has been most cogently and most famously put forward by the American philosopher of science, Larry Laudan.

We can find cases from the history of science of theories that did as well or better than current theories are doing, judged by the best evidential standards of their day. But many of those theories are now known to be false, so we should not think that our best theories are any likelier to be true than the best theories of, say, the 19th century were likely to be true. Laudan is following Kuhn in thinking that our best guide to how science should be done is how science has been done. But Laudan thinks that history shows that realism is unwarranted, since the best standards of actual science permit false theories to thrive. So, the inference from the success of a theory to its truth is blocked by this pessimistic induction.

As Kuhn and his successors have shown, theories like the phlogiston theory of combustion were predictively and explanatorily successful for a long time; they generated normal science; they did all the things that good scientific theories are supposed to do. But right now, we're pretty sure those theories are wrong.

So, what theory in the history of science has been as successful for as long as Newton's physics was, but it turned out to be wrong? The hugely successful wave theory of light, which was adopted by almost all competent physicists in the 19th century is now thought to be wrong. So, the idea is, from the mere fact that a theory looks really good to us, we are in no position to claim that it gets the world right.

In part, this objection can be met by defending some approximation of the traditional view of scientific reduction. If we can show that, at least in many cases, the superseded theory is preserved, gets reduced into the superseding theory, then the history of science no longer looks like a matter of wrong turns and changes of subject; it looks like something that accumulates and makes progress. So, the induction looks less pessimistic, to the extent that we can make such a move, but that involves defending the problematic classical conception of reduction, as we saw a couple of lectures ago.

We'll see this in more detail next time as we try to develop scientific realism, but some promising options for the realist remain. One can narrow one's ambitions and claim that only parts of our best theories are likely to be true, or that only some of our best theories are likely to be true. We need to impose more stringent standards before we make the inference from scientific success to truth.

So, for instance, it could be the case that we have less reason to think our more fundamental theories (basic physics, quantum mechanics, string theory, that sort of stuff) are likely to be true than we have reason to think our less fundamental theories are likely to be true (take plate tectonics as a theory in geology). It's plausible to claim that we have more and better kinds of access to the objects posited by less fundamental theories than we do to the objects posited by more fundamental theories. So, maybe geology is a more secure science, is a safer bet, to infer from success to truth than physics is.

On the other hand, one might think that the lack of solidity at the fundamental level threatens to reverberate throughout science (if physics is wobbly, then everything is wobbly, one might think). It's far from obvious how these arguments should play out. Arguably, even if quantum mechanics is wildly wrong, whatever replaces it had better make room for plate tectonics and molecular biology. Maybe those theories are solid, even if the theories, in some intuitive sense, under them are wobbly.

A somewhat different approach would claim that it's not the entities posited by our best theories about which we should be realists, but something like the mathematical structures involved in the theories. The idea is that the structures of theories change less than the stuff posited by the theories. So, for instance, in the early 19th century, a Frenchman named Sadi Carnot worked out many of the basic ideas of thermodynamics, despite the fact that he mistakenly thought of heat as a kind of fluid, a kind of stuff in things. The way in which kinetic energy gets transmitted between molecules ends up being close enough to the flow of a fluid that the laws of thermodynamics that he formulated are pretty accurate. So, we might be realists about the laws, about the mathematical structures behind the theory, without being realists about what the theory says there is.

The anti-realist will claim that an example like this shows that success is not good evidence of truth, even in good scientific theories, while realists will try to locate the kind of success that leads to truth. We'll distinguish the ontology (the stuff) from the structural features, for instance. The notion of a structure and how a structure can be accurate despite being a structure of the wrong stuff is a tricky problem, however, for realists.

So, we've articulated scientific realism, paying special attention to this tension within the very notion between metaphysical modesty—that the world is "out there" and has a nature—and epistemic presumptuousness—the idea that, despite not getting much direct observational evidence about it, we can get it right.

So, next time, we're going to look at the main arguments in favor of this kind of odd position that tends to have an internal tension within it. We'll try to see what the strong reasons for adopting scientific realism might be, and we'll look at its most serious empiricist competitor, the constructive empiricism of Bas van Fraassen.

Lecture Twenty-Seven
Success, Experience, and Explanation

Scope:

Realists defend their position as the best explanation for the success of science. Anti-realists point to a number of successful-but-false theories in the history of science. Under what conditions, if any, does the success of a theory give us good reason to think that it is true (including in what it says about unobservable reality)? We consider empiricist arguments that the demand for an explanation of the success of science begs the question against anti-realism and constitutes an invitation to metaphysics. We also contrast scientific realism with Bas van Fraassen's constructive empiricism, which combines the semantic claims of realism with the suggestion that scientists shouldn't believe what their theories say about unobservable reality.

Outline

I. *Inference to the best explanation* is the main style of argument for inferring from observable phenomena to unobservable phenomena.

   A. The straightforward argument for realism is often called the "no miracles" argument. The natural sciences have been tremendously successful, and a fairly strong version of realism (the claim that our best scientific theories are at least approximately true, including what they say about unobservable reality) provides the best explanation for this striking fact.

   1. Two kinds of success matter to the "no miracles" argument: predictive and technological.

   2. Given that some kinds of predictive and technological success are cheap, the "no miracles" argument has got to set the bar pretty high if it is to claim that it would be a miracle that science could do what it has done without its theories being at least approximately true.

   3. Even so, the argument runs up against the pessimistic induction argument discussed in the preceding lecture. Predictively accurate and technologically fruitful

theories from the past have been shown to be false. Other generations would have been just as entitled to use the "no miracles" argument, but they would have been wrong; thus, we should not help ourselves to this argument.

  B. The realist needs to require *novel* predictive success before a theory can justifiably be considered true. If a theory explains only data that are already "in," a competing explanation is available for the theory's success, i.e., that it was designed to accommodate the data. Novel predictions preclude this explanation and thereby favor the explanation that the theory works because it is true.

    1. Novelty is tricky to characterize. It's neither a straightforwardly temporal nor a straightforwardly psychological notion.

    2. Even if we confine ourselves to novel predictions, the "no miracles" argument is not unproblematic. The wave theory of light generated precise, surprising, and correct predictions, but it is false.

    3. Another response available to the realist is to argue that the success of prediction is due to a part of the wave theory that was, in fact, correct and that error does not disqualify part of the theory from being true. Only for the highly tested parts of the theory will the realist's explanation of success seem like the best one, and even then, one should admit that it is fallible.

II. Empiricists challenge the whole appeal to inference to the best explanation in the first place. They ask whether the success of scientific theories needs to be explained at all and whether positing the truth of what scientific theories say about unobservables is really the best explanation.

  A. Van Fraassen uses an evolutionary analogy to resist realism. Theories that generate false predictions tend to get discarded, so it comes as no surprise that the theories that remain generate primarily true predictions. But can this deflationary explanation handle the novel predictive successes of science?

  B. Many empiricists consider inference to the best explanation questionable when used within science and even more questionable when used about science. Do we have good reason to think that the world will uphold our explanatory ambitions? Do we have good reason to consider explanatory loveliness a mark of truth?

C. The status of inference to the best explanation is, thus, quite controversial. Realists argue that such inferences are part and parcel of ordinary and scientific rationality, while empiricists emphasize the problems with such inferences and claim that unrestricted demands for explanation tend to lead to metaphysical speculation.

III. Van Fraassen's *constructive empiricism* offers a major empiricist alternative to realism.

   A. Van Fraassen agrees with realists about semantic issues: Scientific theories posit observables in fully meaningful ways. Our theories are committed to the existence of such things as electrons.

   B. But it does not follow that we are or should be committed to the existence of electrons.

   1. All our evidence is observational evidence, and we shouldn't consider ourselves in a position to attain knowledge of unobservable reality.

   2. Thus, we shouldn't believe what our theories say about unobservable reality; at best, we should believe our theories to be *empirically adequate*.

   3. While denying that the distinction between observational and theoretical language can do any philosophical work, van Fraassen maintains there is an important difference between observable objects and unobservable ones. From the viewpoint of science, human beings are a certain kind of measuring device, and our evidence is tied to our size, our senses, and so on.

   4. Van Fraassen permits inductive arguments from observed phenomena to other observable phenomena, and he permits explanatory inferences to observables. It is inference to unobservables (which induction by itself will not get you) about which he is skeptical.

   5. Though van Fraassen does not think that scientists should believe everything their theories say, he does think that they should act as if well-supported theories

are true and should use theories for such purposes as experimental design. We can let ourselves be guided by pictures without believing the pictures.
C. Van Fraassen has shown that the demise of positivism does not mean that empiricism about scientific theories is doomed. But his position is subject to a number of questions.
1. Can the observable/unobservable distinction bear the weight that van Fraassen requires of it? The realist can argue that, despite the fact that we can check only what a theory says about observable reality, we can take methods that we know are reliable with respect to observable reality and apply them to unobservable reality.
2. When we ask why a theory that posits unobservables is predictively accurate and technologically useful, van Fraassen says it is because the theory is empirically adequate. This explanation is likelier than the realist's explanation, but it is very unlovely. Van Fraassen thinks it is no part of science to explain the success of science, but many thinkers find such an explanatory project well motivated.
3. Finally, we can raise some questions about the balance of epistemic modesty and presumptuousness struck by van Fraassen. If we are to be cautious about venturing beyond the observable, why should we not be comparably cautious about venturing beyond the observed? Believing our theories empirically adequate goes enormously beyond the evidence, as the problem of induction shows.

Essential Reading:

Boyd, "On the Current Status of Scientific Realism," in Boyd, Gasper, and Trout, *The Philosophy of Science*, pp.195–222.

Van Fraassen, "Arguments Concerning Scientific Realism," in Curd and Cover, *Philosophy of Science: The Central Issues*, pp. 1064–1087.

Supplementary Reading:

Brown, "Explaining the Success of Science," in Curd and Cover, *Philosophy of Science: The Central Issues*, pp. 1136–1152.

Musgrave, "Realism versus Constructive Empiricism," in Curd and Cover, *Philosophy of Science: The Central Issues*, pp. 1088–1113.

Questions to Consider:

1. Do you think it matters whether one believes a scientific theory or merely accepts it? Does it matter whether one believes some religious doctrine or merely accepts it? Why or why not?
2. To what extent do the major scientific innovations of the last century or so (relativity, quantum mechanics, molecular biology, the rise of psychology, and so on) make scientific realism either harder or easier to defend?

Lecture Twenty-Seven—Transcript
Success, Experience, and Explanation

Last time out, we sketched scientific realism and made at least an initial examination of some of the obstacles the view must confront. We approached this view through the obstacles that it must overcome because it would otherwise seem like a rather uncontroversial claim. Prior to a major dose of philosophy, many people would have thought it completely unproblematic to assent to the claim that scientific theories can reasonably aim at depicting reality correctly.

But against the background of empiricism that dominated 20th-century philosophy of science, it's easy to see that serious problems arise about how data could support claims made about unobservable reality. At this stage of the course, we've dropped the empiricist-inspired semantic worries; we're granting that it's fully meaningful to talk about unobservable reality. But the epistemic issue (we called it epistemic presumptuousness last time) remains: Is it reasonable to think that science can give us knowledge of a reality we do not encounter through observation?

The main argument that's been offered in favor of scientific realism has, like scientific realism itself, a kind of commonsense feel to it—though philosophy, as you know by now, has a way of making commonsense feelings dissipate rather quickly. The argument is a version of an inference to the best explanation—this is an argument form we examined back in Lecture Twelve. Such arguments say that it's reasonable to adopt a hypothesis when it does a good enough job of explaining observations and when no competing hypothesis does a comparably good job. Inference to the best explanation is the main style of argument for moving from observable to unobservable phenomena. Unlike induction, which is sometimes called a *horizontal inference* (it's a "more of the same" inference), inference to the best explanation is a *vertical inference* (we can go from observable phenomena to something behind the phenomena).

This straightforward inference to the best-explanation-style argument for realism is often called the "no miracles" argument. The idea is that the natural sciences have been tremendously successful, and a fairly strong version of realism—roughly, the claim that our best scientific theories are at least approximately true (including what they say about unobservable reality)—is the best explanation for the striking fact that the natural sciences are so successful.

The notion of success at work in this argument calls out for explanation. Lysenko's misguided biology might have served his interests rather well, but that's not the kind of success at issue here. And to the extent that Darwinism caught on in large part because it served the interests of powerful people in the 19th century, that kind of success is also not an argument for the truth of the theory that succeeds in that way.

Two kinds of success do seem to matter for purposes of the "no miracles" argument: predictive success and technological success. But again, the issue needs to be handled with some care. Predictive success can come from luck or from vagueness. Astrology is not without its predictive successes (though generally we might credit those to the vagueness of its predictions), and a lucky guess is a genuine predictive success.

So, the "no miracles" argument has got to set the bar a little bit higher than that if it's going to claim that it would be a miracle that science could do what it's done without its theories being at least approximately true. So, the defender of the argument can appeal to the idea that you find particle physicists but not astrologers testing the implications of their theories with quantitative precision out to 10 decimal places. The idea is how could a theory make a prediction that accurate without the theory being at least approximately true? A similar argument gets applied to technological successes. How can laboratories cook up designer microorganisms if molecular biology is not at least approximately true? And how can you land a satellite in a particular place on Mars if physics is not approximately true?

Understandably, this has been an enormously influential argument, both within philosophy and outside it, but it runs squarely into the argument we called the *pessimistic induction* last time. That argument claims that predictively accurate and technologically fruitful theories from the past have now been shown to be false. So, the idea is that other generations would have been just as entitled to use the "no miracles" argument as we are, but they would have been wrong, so we should not be willing to help ourselves to the "no miracles" argument.

Building on a point that we at least touched on last time, the "pessimistic" argument shows that the realist needs to adopt more demanding criteria of success, or the realists needs to restrict what the "no miracles" argument claims to accomplish, or both.

Taking the first horn of the dilemma, the realist can require novel predictive success before a theory can justifiably be considered at least approximately true of unobservable reality. If a theory only explains data that are already "in," a competing explanation is readily available for that theory's success—namely that the theory was designed to accommodate the data (it's taken the data into account and otherwise might not have gotten it right). Novel predictions preclude this explanation and thereby favor the explanation that the theory works because it's true.

Some philosophers have gone so far as to argue that only novel predictions count as evidence that a theory is true. That's a rather extreme claim. The evidence about how tides and cannonballs and planets work was all in before Newton unified all of these phenomena—but intuitively, we think that his theory should get credit for explaining all of those different kinds of motion as well as the theory did. So, we do think that data that are already in can support a theory.

This restriction of the "no miracles" argument to novel predictions runs into the problem that novelty is rather tricky to characterize. It's not a purely temporal notion because if there's old data that the theory was not able to take in when it was formulated, that serves as novelty in the relevant sense. It looks just as good for the theory because the theory can't have been designed around data of which it was, as it were, unaware.

We don't want to appeal to a kind of psychological notion of surprisingness because that's too person-relative. What we want is some kind of reasonably objective measure of the extent to which a prediction was unexpected from the standpoint of competing theories, and that's a hard notion to formulate.

Even if we confine ourselves to novel predictions and help ourselves to a notion of novelty, the "no miracles" argument is not out of the woods yet. The wave theory of light generated precise, surprising, and correct predictions, but it's false. The bright spot in the middle of the shadow of a disc is unexpected by just about any standard. It's exactly the sort of thing that a theory should get credit for—the wave theory made the prediction, turned out to be true, but the theory is false.

To some extent, this just requires backing away from excessive rhetoric; this is the strand of the response that tones down what the "no miracles" argument can claim for itself. One must admit that, as it were, miracles sometimes happen. A theory can make precise and surprising predictions without being true. That doesn't necessarily undermine the idea that, in general, those sorts of predictive successes are indicators of truth.

Another response along these lines is to argue that the success of the prediction is due to a part of the wave theory that is, in fact, correct, and that error elsewhere in the theory doesn't disqualify part of the theory from counting as true. This revisits some issues going back to our discussion of Quine and holism about theory testing.

The realist needs to figure out some way of determining which parts of a theory have been subjected to suitably various and suitably demanding tests if the realist is going to focus on part of the theory as having been vindicated by its successes. Of course, it's easier to make such determinations about past theories than it is to make them about present ones, so there's at least a kind of anti-realist warning at work here that we shouldn't be too comfortable applying the "no miracles" argument to our present theories. Only for the highly tested parts of a theory will the realist's explanation of success seem like the best one, and even then, as the wave theory of light reminds us, these arguments are fallible.

So, maybe we can grant the realist a limited success in answering the pessimistic induction argument, and perhaps we can grant the realist a limited success in addressing the problem of the underdetermination of theory by data, which—as we saw last time—tries to make hay from the idea that for any set of observations, an infinite number of theories will be compatible with those observations. Let's grant for the sake of argument that, under favorable conditions, a scientific realist can plausibly claim that one theory is better supported by that data than others, even though an infinite number of theories are compatible with that data. Scientific realism is still not out of the woods.

Empiricists can challenge the whole appeal to the strategy of inference to the best explanation in the first place. This kind of an argument can come in two parts. First, the empiricist can ask whether the success of scientific theories needs to be explained at all. In addition, the empiricist can ask whether positing the truth of what scientific theories say about unobservable reality is, in fact, the best explanation of scientific success.

We're going to take the second issue first. We've seen that Bas van Fraassen does not think any context-independent standards for what counts as an explanation or what requires explanation are in place. A good explanation, for van Fraassen, is just a good answer to a "why" question. So, if someone asks about the striking fact that scientific theories are predictively and technological successful, van Fraassen is willing to answer that "why" question. He appeals to a kind of evolutionary analogy in order to do so. We don't need to credit gazelles with any deep understanding of predator/prey relationships in order to explain why they run away from cheetahs; gazelles that didn't run away from cheetahs did not survive and reproduce, so such gazelles are no longer around.

Similarly, scientific theories that generate false predictions tend to get discarded, so it comes as no surprise that the only theories that are still around have mainly generated true predictions. This is an explanation, an answer to the "why" question, but it does not vindicate the approximate truth of what scientific theories say about unobservable reality.

This is a powerful argument against simple versions of the "no miracles" argument because it suggests there's a serious explanation that the "no miracles" argument hasn't ruled out. But the evolutionary analogy might not be able to account for novel predictive success. If gazelles knew enough about the habits of unprecedented predators, we might have to rethink this kind of deflationary explanation of their cognitive skills.

But this leads into the more radical challenge to the whole strategy of inference to the best explanation that empiricists pose for scientific realists. As we saw back in Lecture Twelve when we introduced inference to the best explanation, many empiricists regard this argumentative strategy as questionable when used within science. Given that, it's unsurprising that they consider it even more questionable when it's used about science. Do we have adequate reason for thinking that the world upholds our explanatory ambitions? Inference to the best explanation involves thinking that explanatory loveliness is a mark of truth. In other words, the hypothesis that makes the best sense of things to us (that best unifies our knowledge, for instance), or at least the hypothesis that makes the best sense to us at our best, that hypothesis is likely to be true. The realist is committed to this; the empiricist questions this commitment.

In defense of realism, it can be noted that we appeal to this argumentative strategy, to inference to the best explanation, in lots of ordinary cases, as well as in lots of scientific ones. As we saw, this was the kind of reasoning favored by Sherlock Holmes and pretty much every detective ever since. Until one is hit with a good-sized dose of philosophy, we tend to think that it's reasonable to infer to the truth of a sufficiently good explanation.

On the empiricist side, however, we've looked at various promising but troubling accounts of explanation, and it's not clear that we've seen one that vindicates this strategy of inferring to the best explanation—we are attracted to hypotheses that unify, or that provide a causal mechanism, or that derive from the laws of nature—but appealing to any of these hypotheses also goes well beyond the observational data. That's the source of the worry for empiricists. So, this is another version of the deep tension we've seen between science's ambitions to understand and explain on the one hand, and the caution and modesty with which it approaches the notion of evidence on the other. On the one hand, we're inclined to go beyond observation in order to explain phenomena, and we think our explanations get the world right. On the other hand, we're cautious about going beyond direct observational data.

The same point can be made via the Quinean notion of the underdetermination of theory by data. Suppose we think that the best explanation of certain observations is the truth of our current scientific theory. But if underdetermination holds, it appears that the truth of any of the other theories that fit the data equally well could also serve as an explanation of the observations. How is it that we can narrow down to one the theories that explain the truth of the observation, given that there are an infinite number of them?

It was for reasons like these, as we saw back in Lecture Eighteen when we first started talking about scientific explanation, that hard-core empiricists suspect that explanation tends to lead into metaphysics. They think that we must not posit some fact behind the phenomena to account for every striking regularity (not everything needs to be explained). If we try to explain everything in scientific cases, we'll find ourselves positing unverifiable aether pervading the universe as a way for light waves to propagate, or we'll see an explanation for the number of planets (we saw people explaining the number of planets via an appeal to the number of holes in the head). Pushing the demand for explanation too far leads to metaphysics—at least empiricists fear that that's the case.

The status of inference to the best explanation within science itself is thus controversial. There are powerful considerations in favor of some such inferences, but also powerful reasons for thinking that it's not a general requirement that science explain as much as it possibly can.

Empiricists, then, are not persuaded by the realist appeal to inference to the best explanation to underwrite scientific realism. Someone who is skeptical about the inference from a streak in a cloud chamber to an unobservable electron is not going to be impressed with the much more general and much more abstract inference that we can reason our way to the truth of what our scientific theories, in general, say about unobservable reality from the predictive and technological successes of science.

So, a major empiricist alternative to scientific realism has arisen. This is, as it were, what's going to be left of logical positivism when we drop the semantic doctrines we dropped last time. This view is due to Bas van Fraassen, and it's called *constructive empiricism*.

Van Fraassen agrees with realists about semantic issues: Scientific theories posit unobservables in fully meaningful ways. Our theories are committed to the existence of such things as electrons. Electron talk is not a disguised way of talking about observational predictions. We've rejected that legacy of logical positivism.

But it doesn't follow, says van Fraassen, that we are—or should be—committed to the existence of electrons—our theory is so committed, but we needn't be. Van Fraassen parts company with realists on epistemic issues, though not on semantic ones. All of our evidence, he says, is observational evidence, and this means that we should not think of ourselves as in a position to gain knowledge of unobservable reality.

But since our best theories do make fully meaningful, aspiring-to-truth claims about unobservable reality, we should not believe parts of what our best theories say. That's a peculiar point; we'll return to it shortly.

While van Fraassen agrees with realists that no interesting distinction can be drawn between the parts of our language that concern observable reality and the parts of our language that don't, he insists that there is a crucial difference between unobservable objects and observable objects.

Van Fraassen grants that the distinction is contingent and somewhat vague: It depends on facts about how human beings happen to work. Some things are just barely too small to be seen. Some things might be observable by you, but not by me, because you have better vision than I do. And it's just a contingent fact about our size and our sensory equipment that determines, in the first place, what counts as observable by us.

In full recognition of these facts, van Fraassen nevertheless maintains the epistemic importance of this distinction. From the point of view of science, he says, human beings are a certain kind of measuring or detecting device, and our evidence is tied to our size and our senses. "Observable," he says, is a term that works like "portable." Some things might be movable by you, but not by me, and the Empire State Building might be portable for a sufficiently strong extraterrestrial—but that won't make it portable for human beings like us.

Like "portable," "observable" has its meaning tied to human abilities. Scientific realists tend to think that telescopes and microscopes can be thought of as extending the reach of our senses, extending what counts as observation. Someone like van Fraassen thinks that such devices are sufficiently mediated by theory—by talk about unobservables—that they should be thought of more as inferences than as observations.

And this distinction matters because van Fraassen insists that belief should go only where evidence can lead, and all evidence is observational evidence. So, we should, at most, believe our best scientific theories to be *empirically adequate*—that is, correct in everything they say about observable reality. Again, we're going back to Osiander's Preface to Copernicus's theory of the solar system. The idea is that the job of a scientific theory is to save the phenomena. It is to describe correctly all of the possible as well as actual observations—it's to tell you where you're going to see the planets; it's not to provide a true description of the unobservable reality behind the observations. That's van Fraassen's conception of what makes a scientific theory a good one.

Note that van Fraassen is not a skeptic about scientific inference in general. He permits horizontal inductive arguments from *observed* phenomena to other observable phenomena. And he permits explanatory inferences, inferences to the best explanation, when they confine themselves to observable reality. It's perfectly reasonable, van Fraassen thinks, to infer the existence of an unobserved mouse from observed mouse-like phenomena. It is inference to unobservables (which induction will not, by itself, license) about which van Fraassen is skeptical. To say that a theory is empirically adequate is to go enormously beyond our evidence, and van Fraassen's okay with that. But to say that our theory describes unobservable reality correctly, according to van Fraassen, is to go beyond all the possible evidence, and he thinks realism makes a mistake by going that route.

Van Fraassen, nevertheless, thinks it's okay for scientists to be guided by the pictures of unobservable reality embedded in their theories. You can take, in a certain sense, say, the Copernican picture of the solar system quite seriously—you can take it seriously for designing experiments and doing other things to try to make your theories more empirically adequate.

So (and this is one of the peculiarities of van Fraassen's position), though he agrees with realists that the theory asserts the existence of unobservable objects, van Fraassen agrees with positivists that the epistemic point of making claims about unobservable reality is only to help you make claims about observable reality.

By using your theory to make predictions, to design experiments, et cetera, you are not yet committing yourself to the truth of the theory; it's not yet belief. We can let ourselves be guided by an Earth-centered picture of the solar system when we talk about the sun rising in the morning. We know better, but we can talk that way, and it does no harm as long as we're clear about the conditions under which we endorse being guided by such a picture.

Similarly, in one's capacity as a scientist, you should treat good theories just as if you believe them to be true. You should apply for funding to test them; you should be guided by the picture of unobservable reality presented in the theory. But when you step out of your scientific practice and raise epistemological questions, you're making an assessment of evidence and reasons for belief, and you should only believe your theory to be empirically adequate because that's all your evidence supports.

This raises some interesting questions about the various *attitudes* one might take toward scientific theories: One can believe a scientific theory; one can think it probable that it's true, without believing it; one can adopt the theory as a working hypothesis; one can be uncommitted to the theory, but guided by the picture that the theory presents.

We've seen some views near the various extremes on this spectrum. Popper suggests that scientists are not at all invested in their hypotheses. They treat them as conjectures rather than as beliefs, ready to drop them the moment experience turns against them. Kuhn, on the other hand, emphasizes a kind of dogmatic commitment within normal science and treats paradigms as both widely held and deeply held beliefs that scientists are enormously reluctant to give up.

Though it can appear that way, since van Fraassen allows scientists to "inhabit" theories without believing them, he's drawing on a distinction that is not merely verbal. Van Fraassen thinks it is seriously unscientific to believe where there is no observational evidence. It's not unscientific to act as if we believe, where there's no scientific evidence. That can seem like a quibble, but for van Fraassen, it makes all the difference in the world to a kind of scientific ethic of belief.

Unfortunately, it would take us too far into relatively pure epistemology to pursue this general issue. I am particularly fascinated by this part of philosophy of science. There seem to be some cases in which nothing less than full-fledged belief is appropriate. It is not okay merely to adopt the working hypothesis that you love your spouse, or that all observable phenomena are as if you love your spouse. Sometimes a cautious, distanced, epistemic attitude is appropriate; sometimes a more committed, epistemic attitude is appropriate.

Van Fraassen sees no need for the more committed attitude with respect to unobservable reality. The realist finds van Fraassen's distanced attitude unreasonably tepid about good scientific theories. The realist thinks a scientific theory like, say, a friend has earned more trust than van Fraassen is prepared to offer. Van Fraassen wants to confine the commitment to observable reality.

One's personal, as it were, ethics and aesthetics of belief is likely to play a role in your evaluation of whether van Fraassen's view is a sort-of thoroughly epistemically motivated, sensible view to take towards the distinction between observable and unobservable reality, or whether it's a sort-of bizarrely distanced attitude to take towards good scientific theories.

What is clear is that van Fraassen has shown that the demise of logical positivism does not mean that empiricism about scientific theories is doomed. Nevertheless, his position is subject to a number of questions.

Perhaps the most central one is: Can the distinction between unobservable and observable reality bear the weight that van Fraassen requires of it? From the fact that all of our evidence is observational, it doesn't immediately follow that we can only count as being reliably in touch with observable objects. Realists will argue that, despite the fact that we can only directly check what a theory says about observable reality, we can take methods that we know are reliable with respect to observable reality and apply them to unobservable reality.

The question is whether the burden of proof is on empiricists to give us a reason to think this extension is illegitimate, or on realists to establish the legitimacy of this extension. The philosopher of science, Philip Kitcher, compares this to Galileo's strategy where he first used the telescope to convince his opponents that it was accurate about terrestrial phenomena (phenomena on Earth) and then argued, if it was accurate about terrestrial phenomena, he thought the burden of proof should shift to his opponents to give a reason to think that it wouldn't be accurate when applied to the heavens.

Does van Fraassen draw the line about what is susceptible of explanation in approximately the right place? When we ask why a theory that posits unobservables is predictively accurate and technologically successful, the realist says because it's at least approximately true. Van Fraassen says it's because the theory is empirically adequate. Here is another sort-of ethics of belief, an ethics of explanation point. Van Fraassen's is a likelier explanation than the realists, but it's a much less lovely explanation than the realists. A theory can't be true without being empirically adequate, but a theory can be empirically adequate without being true. So, van Fraassen is going less far beyond the actual evidence than the realist is.

But the unloveliness of the explanation is, for many people, hard to swallow. We want to know why the theory is empirically adequate. Van Fraassen thinks that it is no part of science to have to explain the success of science. The job of science is to succeed—not to succeed and then explain its own success. Van Fraassen is part of an anti-metaphysical tradition going back to, at least, Newton's discussion of gravity. He says, I don't know why gravity works this way; I don't know what gravity is; I'm just describing how observable reality is going to behave.

Only metaphysics, according to van Fraassen, is going to be able to explain the success of science, and metaphysics is a bad idea because of how far beyond observation metaphysics goes. Theories that aren't empirically adequate get rejected, so the theories that aren't rejected are going to tend to be empirically adequate, and that's all the explanation we should want.

Finally, we can raise some questions about the balance between epistemic modesty and epistemic presumption that van Fraassen strikes. Why, if we're to be so cautious about venturing beyond the *observable*, shouldn't we be comparably cautious about venturing beyond what's actually *observed*? Van Fraassen is perfectly prepared to let us generalize our theories to all of observable reality. But believing that our theories are empirically adequate goes enormously beyond the evidence, as the problem of induction shows. So, what exactly is the difference? Van Fraassen thinks it's the difference between what all possible evidence can show and what all possible evidence still can't show (namely, how unobservable reality goes).

But we don't have all possible evidence. Why aren't we restricted by all actual evidence?

So, many observers think that this battle between realism and empiricism ends in a kind of impasse. The realists can't show that our procedures for investigating unobservable reality are reliable without assuming some version of the "no miracles" argument—the claim that empirical and technological success is best explained by the truth or reliability of our theories. But that's the very kind of inference that seems so questionable to empiricists.

In our next lecture, we'll see whether we can at least partially wriggle out of this impasse by appealing to a less-ambitious version of realism's inference to the best explanation strategy.

Lecture Twenty-Eight
Realism and Naturalism

Scope:

These days, scientific realism is generally offered as the best *scientific* explanation of the success of (some) scientific theories. But many empiricists and constructivists object that this amounts to invoking science to testify on its own behalf. What exactly is the claim of circularity here and how damaging is it? Defenders of *naturalistic epistemology* defend a relatively modest conception of justification and emphasize the continuity of philosophy with the sciences. Radical naturalistic epistemologists, such as Quine, have proposed *replacing* epistemology with scientific psychology. We examine moderate and radical philosophical naturalisms and return to the justification of induction as a test case for naturalized epistemologies. We close by asking whether the naturalistic examination of science looks like it will vindicate or disappoint our hopes about scientific reasonableness.

Outline

I. The realist asserts and the empiricist denies that inference to the best explanation can make statements about unobservable reality belief-worthy. In the face of this impasse, many realists have adopted an interesting line of partial retreat. They argue that realism is best defended from within a *naturalistic* approach to philosophy.

   A. Naturalism abandons the project of providing a philosophical justification for science. It gives up on the old, grand conception of philosophy, according to which philosophy can attain *a priori* knowledge through reason alone. But it also gives up on the logical positivists' conception of philosophy as one that tries to achieve valuable results through conceptual analysis alone.

   B. Naturalism is characterized by the rejection of an extra-scientific standpoint from which science can be assessed. For a naturalist, philosophy and science are continuous with one another.

II. A naturalistic approach to realism puts scientific realism forward as the best scientific explanation for the success of science. It no longer attempts a philosophical justification of inference to unobservables.
   A. Scientific realism becomes an empirical hypothesis rather than a philosophical thesis. A naturalized scientific realism takes a scientific look at science and asks whether the successes of science are capable of receiving a scientific explanation. It claims that realism provides the best scientific explanation for the success of science.
   B. The justification offered for realism is that it meets the standards for explanatory inferences that figure in science itself. No attempt is made to address philosophical worries about whether the standards used in science are legitimate.
   C. Like the sociology of science, naturalism involves taking a scientific look at science itself. It involves a scientific examination of the conditions under which scientific practices seem reliable. Naturalists who are realists think that this scientific examination turns out differently than the sociologists believe. They think that the methods of current science can be shown to be reasonably reliable.

III. Two major worries about naturalism arise almost immediately.
   A. Isn't naturalism troublingly circular? Doesn't it amount to judging science by its own standards?
      1. Naturalists are influenced by Kuhn, who suggests that we have no better way of figuring out how science ought to be done than by looking at how it is done.
      2. They are also influenced by Quine's holism, according to which no part of the web of belief stands apart from the rest. In such a picture, there will be no distinctively philosophical or distinctively secure knowledge about how to inquire.
      3. If one were using science to defend the epistemic credentials of science, then the charge of circularity would be well founded. However, that is not what the naturalists are doing. They repudiate the project of justifying science's epistemic credentials in the first place.

4. Rejecting the demand for a philosophical justification must not be confused with having answered it. Science cannot be vindicated by appealing to science. Naturalism refuses to worry about vindicating science.
B. Doesn't naturalism threaten to turn philosophy into some mix of biology and psychology, that is, into the scientific study of how perception, inference, and so on happen? And, in doing so, doesn't it lose sight of the distinction between descriptive and normative questions, between "ises" and "oughts"?
1. Though he later moderated his position, Quine initially defended a strong naturalism along just these lines. He suggested that philosophers get out of the knowledge business. Epistemology should become the study of how science generates such ambitious theories on the basis of such slender inputs.
2. Later philosophers in the naturalistic tradition have been less reductive than Quine was. They think that philosophy can use science to help answer philosophical questions without philosophy thereby becoming part of science. The work of philosophers remains primarily conceptual, but it draws on empirical results.
C. A naturalistic approach to Nelson Goodman's new riddle of induction can serve as an illustration.
1. The naturalist will try to solve questions about legitimate predicates empirically, not conceptually. We should use our best scientific theories to figure out which predicates are legitimately employed in inductive arguments. If the best explanation for the success of a theory is that it employs the right categories, we have some reason to rely on that theory.
2. This approach does not try to address the big epistemological questions about induction. It assumes that such questions have received favorable answers, and it uses science to help answer smaller problems, such as that of figuring out which inductions are better than others.
3. The anti-naturalist will point to the circularity involved in this defense, while the naturalist will ask how we are

supposed to justify anything interesting without using our best theories of the world.

IV. The naturalistic approach does not automatically vindicate current science. Naturalism can threaten, as well as support, our confidence that current science is reliable.

   A. In some respects, naturalism makes the problem of induction even harder to solve, because it raises the problem of obtaining an adequate description of our inductive practices, and that task is very difficult.

   B. Many studies in social psychology appear to show that humans reason badly in certain systematic ways. They violate the basic norms of logic and probability theory.

   C. Evolutionary psychology and evolutionary epistemology suggest that we might be "wired" for some false beliefs about fundamental physics (for example, the impetus theory and a Euclidean geometry of space).

   D. Many sociologists of science think that their empirical work deflates certain myths concerning the rationality and objectivity widely thought to be characteristic of science. A naturalistic approach to science, they think, is incapable of vindicating something like scientific realism.

Essential Reading:

Quine, "Natural Kinds," in Boyd, Gasper, and Trout, *The Philosophy of Science*, pp. 159–170.

Godfrey-Smith, *Theory and Reality: An Introduction to the Philosophy of Science*, chapter 10.

Supplementary Reading:

Kornblith, *Naturalizing Epistemology*.

Questions to Consider:

1. Naturalized epistemology abandons the project of convincing skeptics that science is justified. Do you think that there are many real-life skeptics about scientific justification? How important do you think it is to respond to such skeptics?

2. Can evolutionary epistemology help explain why so much of fundamental physics seems deeply weird to us? Should an evolutionary understanding of human beings alter our conception of what counts as a satisfying explanation, either in physics or in other fields?

Lecture Twenty-Eight—Transcript
Realism and Naturalism

We saw in our last lecture that the debate between scientific realism on the one hand, and empiricism on the other, looks somewhat inconclusive. The realist asserts, and the empiricist denies, that inference to the best explanation can make statements about unobservable reality worthy of scientific belief.

The realist supports this position by appealing to inference to the best explanation arguments within the observable realm, and by appealing to some reasons for not taking the distinction between observable and unobservable reality too seriously.

But the anti-realist (we're mostly concerned with empiricist versions of anti-realism here) argues that inference to the best explanation arguments are not able to carry that much of an evidential burden. The empiricist will appeal to the pessimistic induction argument and the underdetermination of theory by data to suggest that inference to the best explanation goes too far beyond our observational data to be worthy of belief. It's not that the empiricist needs to think that inference to the best explanation arguments are worthless—just that, against this philosophical background, they don't provide an adequate basis on their own for belief. So, the empiricist claims that the realist would have us believe scientific theories on non-evidential grounds, and the empiricist thinks that's a mistake.

Some empiricists think that explanatory loveliness (the extent to which a hypothesis helps us make sense of observation) is never a guide to truth. More moderate empiricists think that explanatory loveliness has some evidential weight within science, and so think that certain kinds of success can license inference to, say, belief in electrons. But they're not persuaded that a global strategy of inference to the best explanation establishes that successful scientific theories correctly describe unobservable reality.

It's simply hard to resolve, at all conclusively, this issue of whether and to what extent it's desirable to tolerate unexplained entities and regularities. Realists tend to think that science should try to explain as much as it can, while empiricists think that the choice ultimately amounts to deciding between saying, on the one hand, "stuff happens, there's no explanation," and on the other hand, going in for metaphysics in order to explain why "stuff happens" (for instance, why science is as successful as it is). Empiricists opt for the "stuff happens" version of that dilemma and claim that we're going to end up, at some point, with unexplained phenomena on our hands, and it's not so bad if that point arrives sooner than the realist thinks it should.

So, that's the impasse that seems to characterize the debate about scientific realism. In the face of this impasse, a number of realists have adopted an interesting line of partial retreat. This retreat is not designed really to bring staunch empiricists along with them, but it does try to, as it were, minimize the controversy over inference to the best explanation. It's independently motivated by other developments in philosophy of science in the second half of the 20th century—in particular, the failure of the logical positivists and their allies to develop a logic or abstract methodology that seemed adequate to scientific induction and other aspects of scientific reasoning. So, at this point, we're bringing in some more material from earlier chapters of our story about evidence and confirmation.

Realism, many philosophers now think, flourishes best within a naturalistic approach to philosophy. Naturalism abandons the project of providing a philosophical justification for science (we'll explore the implications of that momentarily). Naturalism gives up on the idea of any kind of distinctively philosophical knowledge, notably including a priori knowledge (that is, knowledge that does not depend on experience) and—most immediately relevantly—distinctively philosophical principles of epistemic evaluation. So, if we're naturalists, there is no logic or abstract methodology of science; there is no distinctively philosophical justification for science.

Naturalism thus gives up on the old, grand conception of philosophy as the Queen of the Sciences that had so annoyed the positivists. It gives up on the idea that philosophers can attain a priori knowledge through the use of reason alone. But it also, and equally, gives up on the positivists' conception of philosophy as a fact-free discipline—one that tries to achieve valuable results through conceptual reasoning alone, through analysis of language. The logical analysis of science central to Popper and to the logical positivists seemed to many philosophers—by, say, the 1960s or 1970s—no longer to have much left in the tank. They wanted a different way of thinking about the philosophical issues that arise in science.

What's left if we're giving up metaphysics and we're giving up conceptual analysis? Science, pretty much. Naturalism is characterized by the rejection of an extrascientific standpoint from which science can be assessed. Another way of making the very same point is to say that, for a naturalist, philosophy and science are continuous with one another. Philosophy is not prior to science, providing a foundation, nor is it separate from science, a methodologically different discipline handling conceptual rather than factual questions.

We'll examine this view first as it applies to the issue of scientific realism, and then we will return to consider philosophical naturalism more generally.

The naturalistic approach to realism puts scientific realism forward as the best *scientific* explanation for the success of science. It no longer attempts a *philosophical* justification of inference to unobservables. That sounds like a verbal difference. Let's see why it's a difference that makes a difference.

Scientific realism now becomes an empirical hypothesis rather than a philosophical thesis. It's no longer a claim about the nature of justification; it's no longer a claim about inductive logic; it's not a claim about the logical or evidential status of inference to the best explanation. A naturalized scientific realism takes a scientific look at science and asks whether the successes of science are capable of receiving a scientific explanation. The realist claims that realism (which, as we've seen, is, roughly speaking, the hypothesis that successful theories have a reasonable chance of being approximately true, including in what they say about unobservable reality) provides the best scientific explanation for the technological and predictive successes of science.

So, the idea is that, just as electrons are posited to explain the success of certain predictions in physics, scientific realism gets posited to explain the success of certain sciences. Or, to take perhaps a clearer metaphor, just as we can examine the behavior of, say, a group of monkeys (I don't mean to be insulting scientists here), we can look at the monkeys and try to explain their behavior by figuring out which parts of the world they understand well and which parts of the world they understand poorly. We can turn a similar kind of explanatory attention on science itself and try to use science to figure out which parts of science seem to have latched on to reality.

Notice that the only sense of legitimacy appealed to here is derived from scientific practice itself. The claim is that the inference from the success of scientific theories to their at least approximate truth (this inference need not be made about science as a whole; it's made about favorable cases, well-tested theories) meets the standards for explanatory inference within science. It's not directly claimed that it's legitimate; it's claimed that it's scientifically legitimate. It's the same kind of inference involved in good science. So, no external standard and no independent defense of the appropriateness of this explanatory inference is invoked.

Rather than analyzing the notion of justification or evidence (the kind of conceptual a priori work philosophers would have done), the idea is to make a scientific examination of what works, of the conditions under which scientific practices seem to be reliable. This is reminiscent of the approach of the sociologists of science, whose work we examined back in Lecture Seventeen. We've seen that the sociologists of science claim to have examined science and to have no need for the idea that science gets the world—much less the unobservable parts of the world—right. They think a scientific examination of science debunks science's claims to be in touch with reality.

Scientific realists think that this inference turns out differently than the sociologists of science think it does. By the standards of current science, scientific realists—in the naturalistic tradition—think that the methods of current science can be shown to be more or less reliable. And they think that these standards—the standards of our current best science—are the best, if not the only, standards to apply.

We will, in a few minutes, touch on the substantive question here—namely whether scientific realism is, in fact, a good scientific hypothesis (that's roughly the difference between the sociologists of science and the naturalistic philosophers). But it's the methodological issues that are really philosophically striking and interesting here. What makes naturalism so distinctive is that no attempt at all is made to address philosophical worries about whether the standards used in science are themselves legitimate.

Naturalism's utter lack of interest, really, in justifying science raises two major worries almost immediately. The first is, why isn't naturalism troublingly circular? Doesn't it amount to judging science by its own standards? Astrology is likely to do pretty well if we let astrology be judged by the standards of astrology. How is it any different to let science be judged by the standards of science?

An appreciation of some of the historical sources on which naturalism draws can help us see that there's at least something of a difference here. Naturalists are influenced by Kuhn's ideas, which suggest that we have no better grip on how science *ought* to be done than the grip we have on how science *is* done. The failures and limitations of projects like developing a logic of science deepened this sense that we won't be able to say anything very interesting about the "oughts" of science without drawing quite significantly on the "ises" of science. Science is our best intellectual achievement (people in the Kuhnian tradition think), and we've got to look at how it gets done, rather than imposing some external standards on it.

Of course, this is not to suggest that we can read the "oughts" of science too directly off the "ises." It would have been better had Kuhn more explicitly emphasized this point, but even he understands that we must have some basis for selecting which actual scientific practices are taken as typical or exemplary. So the "ises" and "oughts" are intertwined, but in the Kuhnian tradition, the "ises" are primary and the "oughts" are derived from reflection on the "ises" of science.

Naturalists are also influenced by Quine's holism, according to which no part of the web of belief stands apart from the rest. Less metaphorically, nothing can be known a priori, nothing can be known independently of experience, and no belief is immune from the possibility of revision. On such a picture, there will be no distinctively philosophical or distinctively secure knowledge about how to inquire. Inductive logic, deductive logic, any of these principles could be modified in the course of experience.

Just as, for Quine, nothing important turns on whether "Force = mass x acceleration" is a definition (which, broadly speaking, in the tradition, would have been a philosophical claim) or an empirical law instead of a definition (which, broadly speaking, would make it part of physics rather than philosophy), that's not a difference that makes a difference, according to Quine. Nothing important turns on whether a bit of methodological advice is philosophical or empirical. A claim is just a bit of methodological advice about how to run your web of belief.

We saw Quine say that positing atoms and positing Greek gods are attempts to do the same kind of thing—namely to allow us successfully to predict phenomena without having to make needless modifications to the web of belief. The only difference is we tend to think positing atoms does a better job of this than positing the Greek gods. In this vein, Quinean-inspired naturalists develop the idea that philosophy and science are in the same business—namely, the business of efficiently updating the web of belief in the light of experience.

Against this background, we can see that circularity is not quite the issue, though it was a fair worry. If one were using science to defend the epistemic credentials of science, a charge of circularity would be well founded (if we used astrology to argue for the legitimacy of astrology, we're arguing in a circle). But that's not, in fact, what the naturalistic philosophers are doing. They are repudiating the project of justifying science's epistemic credentials in the first place.

Rejecting the demand for a philosophical justification is not to be confused with having answered such a demand. Science cannot be vindicated by appealing to science—that would be to argue in a circle. What the naturalists do is refuse to worry about vindicating science. They treat the epistemic specialness of science roughly like the hypothesis that we're not inhabiting a dream or living in the matrix. They're not trying to defend the reality of the external world; they're not trying to defend the epistemic importance of science—they take it for granted.

A more straightforward issue than scientific realism can provide a helpful illustration here. A kind of subdiscipline called *evolutionary epistemology* appeals to evolution to help explain human knowledge. It asks and answers questions about what our eyes, brains, and such evolved to do well and also what they seem, perhaps, to do poorly. Most versions of evolutionary epistemology offer at least a modest assurance about our basic cognitive abilities. These abilities are too important to the survival of creatures like us for them to be significantly inaccurate. Our ancestors would never have survived and reproduced if too many of their beliefs had been false.

As before, the substantive merits and demerits of this argument are less important to us than the methodological issues raised by these arguments. The arguments of evolutionary epistemology have no chance of convincing somebody not already convinced of the basic reliability of our cognitive faculties, because the truth of evolution—which is presupposed in the explanation—can only be known by using the faculties that are being vindicated.

A proponent of an evolutionary epistemological argument is not trying to establish—from minimal presuppositions—the reliability of our cognitive faculties. That's presupposed in the argument. The approach assumes that the scientific picture of the world is more or less accurate and tries to see how philosophical questions look from such a standpoint. The idea is to use the results of science to help answer philosophical questions.

This brings us to the second major worry about philosophical naturalism. Doesn't it threaten to turn philosophy into some kind of mix of biology, and psychology, or maybe other parts of science? Doesn't it threaten to turn philosophy into the scientific study of how perception and inference actually happen? Where are the "oughts"? It's all "ises." And in doing so, doesn't naturalism lose sight of the distinction between descriptive and normative questions?

The precursors of this movement are once again illuminating. We've already seen Kuhn's skepticism about the usefulness of separating "is" questions from "ought" questions. So, the distinction between the normative and the descriptive is not one that Kuhn thinks will bear any interesting weight.

Quine took a pretty interesting position here. He initially defended a very strong naturalism along just these lines. He suggested in the late 1960s that philosophers get out of the knowledge business. Epistemology should become the study of how science generates such ambitious theories on the basis of such modest inputs, and that's more or less a branch of psychology.

Later philosophers in the naturalist tradition have not been as reductive as Quine was (perhaps because we'd like to keep our jobs). Philosophers have seen themselves asking traditional philosophical questions from within a framework that starts by assuming a roughly scientific picture of the world.

Philosophers raise questions about how well certain scientific pictures of the world fit together with others. So, for instance, physiological psychologists tend to think that human attraction is entirely explained by pheromones, while social psychologists tend to think it's entirely explained by certain kinds of affinity in background, interests, perception of oneself and others. Are these competing explanatory pictures? Do they support one another? Which of them is more basic? That's a kind of philosophical question about the status of these scientific explanations.

Similarly, philosophers raise questions about the relationship between the picture of the world offered by various sciences, and that of commonsense. As we'll see in our next-to-the-last lecture, philosophers wonder whether psychology actually will allow us to retain a sense of ourselves as persons with beliefs, plans, and desires. Some developments in psychology suggest that those categories are misguided. And insofar as there's a tension between, say, our thinking of ourselves as free versus our thinking of ourselves as determined scientific objects, how are we to decide whether it's the scientific or the commonsense picture of ourselves in our world that should yield?

Most naturalistic philosophers still think that philosophy has a distinctive role in addressing normative questions. Philosophers generally worry a lot more than scientists do about what goals we have and how well they fit together. So, naturalized epistemologists try to help us get clear about the different things we might mean by a term like "knowledge" and try to help us reflect, using the results of science, on which of these terms has a useful role to play for assessing our belief-forming practices. Cognitive science tells us how we form beliefs, while philosophy helps us ask and answer questions, using cognitive science, about how to assess and improve our belief-forming practices.

In all of these ways, then, philosophy can use science to answer philosophical questions without thereby replacing itself with science. The work of philosophers remains distinctively conceptual and distinctively normative, but it draws unashamedly on empirical results.

A solution to Nelson Goodman's new riddle of induction (better known as the "grue problem") can be used to illustrate how naturalism might work in practice. You may recall that our observations of emeralds seemed to support indefinitely many inductive conclusions. All observed emeralds are green, but they are also "grue" and "gred" and "grack"—you name it.

Some traditional philosophical approaches tried to banish these "gruesome" predicates based on their logical form, and there's still work being done in this area. Even if you're a naturalist, you needn't think that all philosophical work has to be naturalistic.

But the naturalistic defense itself claims that it's an empirical—not a conceptual—matter as to which inductive generalizations, if any, are justified. We won't find a philosophical notion of a real property that will get rid of "grue."

But if we can work out an argument supporting the claim that "green" is a scientifically legitimate predicate in a way that "grue" isn't, that provides a reason—within this assumed naturalistic framework—for favoring the hypothesis that emeralds are green over the hypothesis that they're "grue." Notice that fitting with the data can't be the notion of scientific legitimacy here. "Grue" fits with the data every bit as well as "green" does. What we're looking for is some kind of support from an assumed scientific theory, not just from observation. If science tells us that green things have something important in common that "grue" things lack, then we have a reason for preferring the green hypothesis to the grue hypothesis. It's not, by any means, clear that science says any such thing, by the way.

More generally, we use our best scientific theories to figure out which predicates are legitimately employed in inductive arguments. If the best explanation, scientifically speaking, for the success of certain theories is that they employ the right categories, then those categories can reliably (though, of course, not infallibly) figure in inductive arguments.

This naturalistic approach refuses to take on Hume's general challenge to induction, and by doing so, it gets the ability to appeal to science without having to justify that appeal, and so it uses science to tell us which inductive arguments are likely to work. On this view, science can't help us answer the biggest epistemological questions (like whether the future will resemble the past), but if you assume—in some small respects—favorable answers to such questions, science can then help us answer pretty big epistemological questions, like which inductions look more reasonable than others.

The anti-naturalist will point out that there's a circularity involved here—that we're appealing to science in order to vindicate some scientific claims. The naturalist is going to reply by asking how we're supposed to justify any interesting claims about the world without using our best theories of the world. So, the naturalist will be unapologetic about the appeal to science.

We've actually seen a somewhat similar naturalistic move when we examined the causal or historical approach to meaning and reference. The idea there was that our best scientific theories determine at least big parts of the meaning of certain terms. It's science that tells us what the structure is that makes something a "whale" and which thereby determines which creatures fall under the term "whale" when somebody points at a whale and says, "Wow, what a big fish" (even though whales aren't fish, they're talking about whales, and science tells us which creatures are being talked about).

If it were automatic that naturalism would vindicate current science, then we'd more or less be helping ourselves feel better about our inability to answer Hume's challenge, our inability to develop a scientific logic. We'd just be patting ourselves on the back by assuming that our best scientific theories are on to something.

But fortunately, it's a lot more interesting than that. Just as an observation statement couched in the language of a theory can nevertheless have the power to undermine the theory, a naturalistic approach to inquiry can threaten, as well as support, our confidence that current science is reliable. It's for this reason that naturalistic philosophy preserves much of the liveliness of classic philosophy of science, much of the sense of asking fundamental questions about how inquiry ought to be conducted.

So, for instance, it's not clear to what extent assuming naturalism really gets us very far towards solving even modest versions of the problem of induction. If we're to muster a scientific defense of our inductive practices, we at least need to begin from a scientifically defensible description of those practices.

But even that modest-sounding task of describing which inductive arguments we use in science is quite beyond us so far because we use a kind of weird mix of inductive arguments in the narrow sense ("more of the same" inferences); we use some explanatory inferences; we appeal to simplicity, to fruitfulness, to scope, to variety of evidence. Logically equivalent hypotheses, like those that figure in the Raven Paradox, don't seem equivalent to us.

As Kuhn pointed out, our various inductive norms often compete with each other. It's not clear what part of this package the success of science is supposed to vindicate, or how it can do so. A naturalistic approach to the problem of scientific induction, it's not clear whether it supports or undermines our confidence in our inductive judgments. When we turn science on our inductive practices, to a certain extent, we see a mess.

On a related note, many studies in social psychology appear to show that humans reason quite badly in certain systematic ways; they violate basic norms of logic and probability theory.

Here's an example due to the social psychologists Amos Tversky and Daniel Kahneman. They gave subjects some background information about a young woman. The background information says that she's very bright, outspoken, and deeply concerned with issues about discrimination and social justice. Then they asked their subjects to rank a number of statements about which descriptions are most likely correctly to apply to this young woman. About 85 percent or so of subjects thought it more likely that the woman is a bank teller who is active in the feminist movement than that she is a bank teller.

There are a bunch of different replies you could have given; some of them are nearly irrelevant. This one can't possibly be right. The response given by 85 percent of people is necessarily wrong. There is no conceivable way that it's more likely for something to have both properties A and B than it is for it to have property A alone—that's not possible. But this is a very high percentage of people who gave such an answer, and it's a surprisingly robust result when you modify the experiment in certain respects.

So, naturalized epistemology and philosophy of science can thus raise, in a scientifically informed way, issues about which norms do govern our reasoning and issues about what norms should govern our reasoning, and how we're supposed to figure out what the relevant norms are, given that there's some reason to think each of us is tempted to reason badly in pretty systematic ways. It's far from true that the way we do think lines up unproblematically with the way that we ought to think (or at least with the way that we think we ought to think). Even assuming that science is more or less an epistemically special practice does not vindicate our most basic reasoning practices.

Similarly, as we've seen, evolutionary psychology or evolutionary epistemology generally suggests that our intellectual and perceptual faculties have to be reasonably well attuned to our environment, or else our species would not have survived. On the other hand, it also suggests that we might well be "wired" for some false beliefs about fundamental physics. Some experiments suggest that human beings have a kind of intuitive physics, which is Aristotelian, a kind of impetus theory. It says that if you were to put your arm in a circular motion and let a ball go, it would continue in an arc rather than, as Newton correctly says, off in a straight line tangential to the circle. Similarly, we might be wired for a kind of Euclidean geometry of space, which is probably not the correct geometry of space.

And our difficulties in getting our heads around relativity theory, quantum mechanics, string theory, all of these suggest that we might not be well suited to understanding some parts of the universe. Science might suggest that parts of our commonsense world view are—for evolutionary reasons—likely to be reliable, but other parts are likely to be quite unreliable, irrelevant to our survival, and perhaps beliefs that are relevant to our survival impede our ability to understand other aspects of the world.

And as we noted in our earlier discussion, many sociologists of science think that their empirical work deflates certain myths concerning the rationality and objectivity widely thought to be characteristic of science. A naturalistic approach to science, in their view, is incapable of vindicating something like scientific realism because the categories of truth and objectivity are not properly scientific; they don't figure in legitimate scientific explanations. Naturalistic philosophy of science can thus seem like kind of an awkward hybrid to such people, of the old philosophical aspirations to ground science and the newer, more scientific approaches to science.

With these challenges in mind, we're going to turn in our next lecture to see to what extent a naturalized approach to the philosophy of science vindicates science's special place in our intellectual life. If we examine science in the way that the sociologists suggest, turn a scientific eye to how science works, do we vindicate science as epistemically special, or do we see through some of its pretensions?

Lecture Twenty-Nine
Values and Objectivity

Scope:

Recent work in naturalistic epistemology has turned to the social structure of science. This work has been much friendlier to traditional ideals of objectivity than was the strong program in the sociology of knowledge. But the ideal of objectivity need not be thought of as value-free or disinterested. This lecture examines the values, motives, and incentives that animate science and scientists. To what extent are these values cognitive and to what extent is it a problem if they're not? Might the social structure of science generate objective results even if individual scientists are motivated by the pursuit of recognition, money, or tenure? In what ways might the social organization of science be changed in order to increase objectivity? Who should get to participate in the formation of a scientific "consensus" and why? To what extent can the need for scientific expertise be reconciled with the democratic ideal of citizen involvement in important decisions?

Outline

I. Social factors—money, prestige, political and economic interests, and so on—have often loomed large in the actual practice of science. It has often been implicitly assumed that these social aspects compete with norms of rationality and objectivity that also figure in scientific conduct.

  A. For the positivists, social factors tend to distort the objectivity that would otherwise result from the application of the scientific method (at least within the context of justification).

  B. For many of the sociologists of science, appeals to evidence and logic mask the operation of non-evidential interests and biases that constitute the real explanation of scientific conduct.

  C. We have seen a position between these two views in the work of Kuhn, for whom social aspects of the organization of science can aid, rather than impede, the rationality of science.

II. Recent work in naturalized epistemology and philosophy of science has followed Kuhn in developing a position according to which social and epistemic norms can cooperate, rather than compete. It has followed the sociologists of science in thinking that even normal science is significantly governed by nonepistemic factors, and it has followed the logical positivists and others in thinking that science is, for the most part, epistemically special.

   A. It is clear that it can be disastrous for science to be driven by ideology, but it is not clear that ideology need be epistemically harmful to science.

   1. Suppose a classic Marxist critique of science to be entirely correct: Science serves the interests of industrial capitalism. It is plausible that such ideology-driven science would be highly reliable, because industrial capitalism values accurate information about the empirical world.

   2. Such science could count as objective without being disinterested.

   3. This kind of "invisible hand" defense of scientific objectivity will be subject to very severe restrictions, and it does not show that ideology won't lead to scientific distortion. But it's worth noting that ideology doesn't automatically lead to such distortion.

   B. It can also be argued that the reward structure of science, on the whole, has epistemically salutary effects.

   1. Scientists are rewarded (with prestige, among other things) for having their ideas cited and used. This encourages finding original results and making one's ideas available to others.

   2. Because scientists rely on the ideas of other scientists, the reward system creates some pressure toward testing and replicating the results of others. Ideas are tested through a kind of cooperation and through a kind of competition.

   3. The reward system has some tendency to promote a healthy distribution of scientific labor. If many people are pursuing the most developed research project, it can be rational for other scientists to pursue alternatives.

4. Although the ordinary self-interest of individuals can lead to a community that functions in a more or less disinterested, inquiring manner, the increasing role of money in science and the recent upsurge in corporate sponsorship of research complicate this model considerably.

III. Ideology and other sources of idiosyncrasy certainly have exerted embarrassing influence on science in the past, and this observation raises important issues about how objectivity can be cultivated and increased in science.
  A. Individual scientists sometimes evaluate hypotheses partially on the basis of non-evidential factors.
    1. Some such factors are politically significant (such as gender, class, race, or nationality), while others arguably stem from considerations as apolitical as birth order.
    2. One might expect these non-evidential factors to figure differently in some sciences than in others. Assumptions about gender seem to have crept into primatology but don't seem like much of a worry in theoretical physics. An individual scientist's aesthetic sense might loom large in theoretical physics, however.
  B. Such protection from distortion and idiosyncrasy as science possesses rests less on finding impartial judges than on structures that bring a range of relevant critical perspectives to bear on ideas and their applications.
  C. This raises questions about the diversity, in terms of gender, age, birth order, politics, style of intellectual training, and so on, of a given field. Ideally, it seems, you would want as much variety as you could get in order to bring effective criticism in the field.
  D. The objectivity of a given scientific field is increased by its openness to criticism. Does the field have good conferences and journals? A number of mechanisms can operate to prevent criticism from being as effective as it might be.
  E. But a version of the "white noise" problem looms large here. Diversity of background and opinion has costs as well as benefits. Requiring evolutionary biologists to take creation scientists seriously might have some tendency to increase the

objectivity of the discipline, but it's not clear that it's worth the opportunity costs of doing so.

IV. Questions about values and the social structure of science loom even larger when we turn our attention to science's role in society at large.

   A. Privately funded science would seem legitimately to serve narrower interests than publicly funded science, but it figures in the public sector when it makes a claim to guide policy or to reveal the truth about something.

   B. To what extent do scientists have an obligation to reflect on the likely uses of their research? Can one make the argument that the pursuit of knowledge is justified in itself and that the moral consequences should be left to those who apply the research? Much turns on the extent to which benefits and harms of a research project are reasonably foreseeable.

   C. Issues also arise about how scientists obtain their data. In the United States, if people participate in a medical study, they are owed the highest standard of care. Is it permissible to run studies in other countries for the purpose of avoiding this expensive burden?

   1. On the one hand, the researchers seem to be using people as guinea pigs, taking advantage of already significant inequalities.

   2. On the other hand, they might well be offering their research subjects better medical care than they would otherwise get. We leave these sorts of issues to ethicists.

   D. Finally, we note difficulties about scientific decision-making. Nonscientists must rely on scientists to ascertain the scientific significance of such a proposal as the superconducting supercollider. Who should decide whether a supercollider gets built, and how should such decisions be made?

Essential Reading:

Railton, "Marx and the Objectivity of Science," in Boyd, Gasper, and Trout, *The Philosophy of Science*, pp. 763–773.

Longino, "Values and Objectivity," in Curd and Cover, *Philosophy of Science: The Central Issues*, pp. 170–191.

Supplementary Reading:

Godfrey-Smith, *Theory and Reality: An Introduction to the Philosophy of Science*, chapter 11.

Kitcher, *Science, Truth, and Democracy*.

Questions to Consider:
1. Which parts of science seem most and least ideologically driven to you? In the relatively ideological parts of science, to what extent does the presence of ideology undermine objectivity?
2. Which sciences seem to you to strike the best balance between Popperian openness to criticism and Kuhnian consensus about standards and procedures?

Lecture Twenty-Nine—Transcript
Values and Objectivity

We saw last time that philosophers who adopt a naturalistic approach are willing to allow results of science to help settle questions about science. In this respect, philosophers are following a path that was suggested by Kuhn and followed by others, away from the idea of philosophy as a radically unempirical discipline, primarily as a conceptual discipline.

This also brings philosophy closer to the approach of the sociologists of science that we saw back in Lecture Seventeen. They rejected conceptual analysis, the idea of logic of science, and the focus on normative questions construed as abstract or unrealistic in their view, and ungrounded in empirical work. So, the sociologists put themselves forward as replacing conceptual analysis as the way to approach normative questions about science.

Naturalistically inclined philosophers have generally thought that the sociologists overreacted against the stringency of logical positivism. They've generally held that the sociologists were wrong to deny the relevance of notions like truth or objectivity to a naturalistic explanation of science and its success. In this lecture, we examine naturalistic approaches to the social structure of science and their consequences, approaches that at least stand some chance of vindicating some of science's epistemic ambitions.

But we should realize at the outset that it seems undeniable that social factors—money, prestige, and political and economic interests—have often loomed large in the actual practice of science. The sociologists weren't wrong to insist on that. It's often been implicitly assumed that these social aspects compete with norms of rationality and objectivity that also figure in explanations of scientific conduct. Few people come out and claim that an explanation in terms of social factors precludes one in terms of epistemic factors, or vice versa. But both those who favor the epistemic explanations of scientific behavior, and those who favor social explanations, have tended to act as if their kind of explanation is the only one that really matters.

So, as we saw, the positivists tended to think that social factors distort the objectivity that would otherwise result from the application, in a disinterested way, of the scientific method (at least within the context of justification; psychological and social factors could operate within the context of discovery). This can count as one extreme for our purposes in this lecture.

For many of the sociologists of science, on the other hand, appeals to evidence and logic function mainly to mask the operation of non-evidential interests and biases. It's these interests and biases that constitute the real explanation of scientific conduct. This can serve as the other extreme.

We've also seen a position in-between these two views in the work of Thomas Kuhn—for whom social aspects of the organization of science can aid, rather than impede, the rationality of science. Kuhn held, for instance, that the relatively permissive norms that govern during scientific crises, combined with social and psychological idiosyncrasies of individual scientists, leads to a beneficial distribution of workers and resources. It's a mechanism by which science manages to have some people pursuing the old paradigm and others trying to develop new ones, which is exactly what you'd want during a time of crisis.

Recent work in naturalized epistemology and philosophy of science has more or less followed Kuhn in developing a position according to which the social and epistemic norms can cooperate, rather than compete. But they've followed the sociologists of science in thinking that even normal science is significantly governed by non-epistemic factors, while also following the logical positivists and some of their fellow travelers in thinking that science is, nevertheless—for the most part—epistemically distinctive. That's a peculiar combination of views, and our task is to try to see how they hang together.

Everyone will admit that it can be disastrous for science to be driven by ideology (Lysenkoist biology under Stalin is perhaps the clearest example). Even somebody who thinks that all science is driven by ideology can grant that some ways in which science is driven by ideology can be particularly harmful. It's much less clear that ideology is necessarily epistemically harmful to science.

Let's consider an argument due to a philosopher of science named Peter Railton, who I should admit was one of my teachers, so perhaps I'll be too easy on this argument. Suppose, Railton says, that an old-fashioned Marxist critique of science was entirely on the money: Science relentlessly serves the interests of industrial capitalism. What could be more ideologically driven than that?

It's, nevertheless, plausible that such ideology-driven science would be highly reliable. Why? Industrial capitalism highly values accurate information about the empirical world. Capitalism provides strong incentives for individuals to formulate new hypotheses about how the natural world is structured and how objects will behave. It provides strong monetary incentives for testing these ideas and—under conditions of copyright, and intellectual property, and stuff like that—for propagating those ideas.

A term like "objectivity" is a notoriously tricky term. It often means something like "disinterestedness," and on this Marxist hypothesis, natural science would be anything but disinterested. But a process is also sometimes called "objective" when it is mainly determined by the objects it says it's about, whether or not agendas or biases play a role in that determination. And it's plausible that the biases and incentives of industrial capitalism motivate beliefs that are determined by their objects. One is biased to get the world right, not out of some disinterested desire for knowledge, but nevertheless, the bias serves to help one get the world right.

As Railton would be the first to insist, the story is not as simple as I've made it sound. It's only going to be plausible where there's the right sort of feedback mechanisms with nature so that, if my hypothesis is wrong, somebody else will have the opportunity and, plausibly, the incentive to correct it.

One might well suspect that such feedback mechanisms are going to be lacking, at least to some extent, in a discipline like economics (it's not clear that there's a causal interaction between us and the objects of economic inquiry). One might further expect (at least if we continue assuming a Marxist perspective for the sake of argument) that ideology will provide less incentive to have our ideas produced or caused by the objects of inquiry in economics. According to the Marxist, anyway, capitalism has reasons to hide economic truths from itself, though it doesn't have reason to hide, say, geologic truths from itself. So, this case will work quite differently from cases in the natural sciences.

Similarly—to some extent, anyway—Railton's invisible hand argument about the epistemic status of science will be subject to the same kind of limitations as more classic versions like Adam Smith's famous invisible hand argument. Too much power concentrated in the wrong hands can interfere with the desirable mechanisms of the system. If a company or individual can effectively squelch competition, the incentive structure starts to wobble (which is why government regulation against monopolies is provided in classic capitalism).

Perhaps such examples as corporate-sponsored science aimed to cast doubt on global warming suggest that, after all, ideology does lead to distortion in the long run (and we could multiply examples by looking at, say, biomedical research). So, it would be too sanguine to claim that ideology and objectivity are never in tension with each other when science takes place within a broadly capitalist economy.

But the weaker claim is the really noteworthy one here—namely, that it's not automatically the case that there's a tension between ideology and objectivity. Of course, this point doesn't depend on facts about capitalism and Marxism. For a naturalistically inclined philosopher, it's an empirical question to what extent a given kind of scientific bias distorts. Some biases that are not at all epistemic can be epistemically beneficial.

Similarly, ideology is not the only non-epistemic source of potentially beneficial epistemic effects. One could argue that the reward structure of science, on the whole, has epistemically welcome results. Scientists get rewarded with things like prestige for having their ideas cited and used. This encourages finding original results, so that you can publish them and get prestige, and also it includes making one's ideas available to others. Those seem like good things.

Since others use one's ideas by relying on them in their own work, it creates some (albeit quite imperfect) pressure towards testing and replicating the results of others. Ideas can get tested both through a kind of cooperation (those who want to rely on one's ideas might test them either before or during their own work) and also through a kind of competition (those whose work is threatened by one's work in various ways have an incentive to see whether the result was properly arrived at, or whether, say, your experiment was badly constructed in some way).

We don't want to sound Pollyannaish about this. The reward system also tends to encourage things like plagiarism and fraud; it also encourages protecting oneself against such abuses, so it's plausible to think that such problems only occur occasionally.

The reward system of science also has a tendency to promote a reasonably healthy distribution of scientific labor. If a great many people are pursuing the most developed and the most promising research project, it can be rational for other scientists to pursue alternative projects. This is because rewards and credits are—at least to a first approximation—inversely proportional to the number of people working on a project. The more people who produce a successful result, the more the credit has to be shared. So, it might be reasonable—for purely selfish reasons—to devote part of one's career to a long-shot project because the payoff is bigger, though perhaps less likely.

This is an application to normal science of a point Kuhn had made with respect to revolutionary science. Here, we're not relying—as Kuhn had—on randomizing factors (as Kuhn appealed to individual idiosyncrasies to explain distribution of scientific labor across revolutions). The explanation here is more structural, more built into the rewards system. But the picture is similar in that choices at the individual level are supposed to make for a good distribution of labor at the level of the science as a whole.

To some extent, anyway, self-interest at the level of individuals can thus lead to a community that functions in a more or less disinterested, inquiring manner. The desire to have one's work shared encourages sharing of information, but competitive factors and a desire not to have one's work dismissed, because it relied on shoddy work by other people, encourages criticism.

Once again, this may be too comfortable a picture, at least in certain respects. Money and similar incentives are starting to play a much larger role in science than they have in the past. Issues like the corporate sponsorship of research complicate the sort-of rewards system model considerably. It's commonly claimed, for instance, that 10 percent of biomedical resources go toward diseases that account for about 90 percent of human suffering. A lot more money is spent on hair loss and erectile dysfunction than on tuberculosis.

Note that there is no way of distributing resources that's going to count as free of some kind of interest or agenda. The humanitarian agenda in biomedicine isn't a purely scientific one, whatever exactly we might have in mind with a phrase like a "purely scientific agenda." There are theoretical questions that matter to science without having any clear practical payoff, but we shouldn't confuse that with the idea that something is value-free. The talk of mattering suggests that even when science is free of practical agendas, it's not a value-free enterprise. We should, I suggest, try to clarify and defend the values that animate our research, not pretend that our research is animated by no values at all. This is a common—though not an uncontroversial—opinion.

So, it's plausible to argue that there's something epistemically benign about the reward structure of science, but on the other hand, it hardly seems unimprovable. This is another way in which a naturalistic approach to philosophy of science can raise normative questions. We could, perhaps, set up a reward system that makes science more objective, more efficient.

So far, we've looked at the non-epistemic interests built into the reward structure of science. But this not the kind of non-epistemic interest that gets people's attention—that gets them exercised. The high level of volume attained during the Science Wars of the '90s concerned such matters as the role of political (in a broad sense of "political") values in science, so let's return to the issues of ideology. It's these kinds of non-epistemic factors that get people worked up about whether science is objective or not.

Just about anybody will have to admit that prominent scientific work has sometimes been rather embarrassingly influenced by political or religious ideas and by conceptions of gender and race. To take just one example, some really impressively tortured arguments were invoked about skull and brain size during the 19th century to make sure that it turned out that males of European origin came out ahead on whatever measurement would most matter (because it had to be correlated with intelligence somehow), and what the physical basis of intelligence was could vary, as long as it ended up favoring the right people. It started off being absolute mass of the brain, and then it was body ratio, because it couldn't end up that women or people of non-European origin would finish ahead on some ratio that mattered.

Along with the political kinds of considerations that sometimes affect the judgment of scientists, we should note that things like birth order (at least according to an influential recent book) shape the intellectual personalities of scientists in striking ways. It suggested that first-born children prefer sort-of order, and second-born children are much more sort-of adventurous, attracted to different hypotheses than their older siblings are.

These sorts of factors presumably figure differently in different sciences. Certain masculinist assumptions are widely cited in the biology literature, perhaps most famously in the Sleeping Beauty conception of conception—according to which the sperm cell is struggling heroically upstream, looking for an egg to fertilize, while the egg is just sort of hanging out, waiting for something to happen. This does not accurately represent the biological facts, but it's a way that at least I learned back in 10th grade how human reproduction worked. Similarly, within primatology, some feminists have worried that the alleged sexual passivity of female primates is much exaggerated by some observers.

In physics, one would expect to find less directly political cases of influence, though the case has been made that women tend to see things as more interconnected than men do, and so take different hypotheses more seriously than men would. Some feminist scholars, on the other hand, think that this is little more than a stereotype of women—and women scientists, in particular—that oughtn't to be dignified by a place in the discussion of scientific objectivity.

One might also expect to find a fairly large role that is played in physics by aesthetic considerations of simplicity and explanatory elegance. So, matters of intellectual style of a sort emphasized by the birth-order hypothesis about scientific personalities might matter a good bit more in physics than, say, in biology.

The nature and significance of these cases needs to get examined on an individual basis. Our interest in this lecture is in the strengths and weaknesses of the social structure of science at handling potentially distorting factors. We don't have time to look into, in a given case, what sorts of factors there are and how distorting they are.

Individual scientists are often capable of great objectivity, but few of us are aware of all of our major biases. Such protection from distortion as science possesses really rests less on finding impartial judges of evidence than on bringing a range of critical perspectives to bear on how people process the evidence.

Ideally, we'd want as wide a range of perspectives as we can, able to bring relevant criticism to work on how evidence gets interpreted. This is a point that had, a while ago, been emphasized by Feyerabend—though most people since Feyerabend approach it less dramatically than he did (you can't approach things more dramatically than Feyerabend had). The idea is that the "use and give credit" system can help sort out idiosyncratic interpretations of data—misleading descriptions of experimental results—in the service of political agenda, or some unexamined bias, or something like that.

This immediately raises questions about the diversity—in terms of gender, age, birth order, politics, and style of intellectual training that characterizes the scientists in a given field. Again, ideally, it seems you'd want as much variety as you can get, at least with respect to any features that seem relevant to the field in question. So, gender might loom larger if you're trying to put together a research team investigating the sexual behavior of chimpanzees than it might if you're doing string theory. Again, even here, you have to wonder whether women, as such, are supposed to have anything deep in common, or whether this is a kind of stereotype. These are hard questions.

The point of this approach, though, is not so much about head counting. It's about how effectively criticism can be brought to bear in a field. This is an important issue about scientific objectivity. It crops up in many ways, and the consensus seems to be that a naturalistic approach shows that science does pretty well, but it could certainly do better. Science could be more objective than it, in fact, seems to be.

The questions to ask about objectivity within a field tend to have a kind of Popperian flavor; they're about the ways in which a field genuinely—not just nominally—opens itself up to criticism. Does the field have good conferences and journals (where "good" here means that ideas are really exchanged rather than self-congratulatory speeches given)? Are established people taken too seriously and young scientists not seriously enough (or perhaps vice versa, as sometimes happens when a field gets taken over by a new trend)? Is criticism really valued within the scientific community? It's often not prestigious to be a good editor of a journal or a good referee of journal papers, but some have claimed that that's as important as original research is to the health of a field (though that's not necessarily what deans look at when they're looking to hand out promotions and raises).

We could also worry about whether there's a "good ol' boy" or "good ol' girl" network, the members of which manage (and I'm not suggesting that this often happens intentionally) to kind of credential each other by publishing articles, writing letters of support, et cetera, while again, unintentionally, insulating their shared views from criticism? They're appealing to their own standards of good work and not listening to other people, and they have enough institutional heft to make that a self-sustaining kind of program (that sometimes happens). Do the practitioners in a field generally conduct themselves in a way that discourages those with different training or different approaches from joining or remaining in the field?

One could also worry about whether the standards of rigor within a field are, in fact, deeply connected to producing good work, or are some of them gratuitous (as we sometimes find in philosophy, where you have to show how much logic you know, even if it's sort of incidental to making your point clearly). You don't want standards of rigor that are just designed to show that you're smart or tough. That excludes people who might have something valuable to contribute.

Given all of these issues, along with the more straightforwardly political ones, almost any field or subfield will fall short of some ideal of perfect objectivity. But there's also a major worry from the other side—more, as it were, from the Kuhnian wing of philosophy of science than from the Popperian wing. As we saw, Kuhn emphasized the importance of shared belief, of a shared paradigm, to the success of normal science. By trying too hard to increase objectivity, we raise a kind of "white noise" problem. Diversity of background and opinion is great in terms of bringing criticism to bear on views, but it has costs as well as benefits—it gets in the way of actually doing research.

How much attention are evolutionary biologists really supposed to pay to creation scientists? We saw that even if we can't settle the demarcation problem, there are limited resources available for scientific research and discussion. A given science has rough-and-ready gatekeeping criteria, but there are legitimate worries that these can be too permissive, that they can let in fringe views that are a distraction from getting on with the business of science, and that they can be too restrictive—that they decrease scientific objectivity needlessly by silencing voices that could productively be added to the conversation.

So, it's worth asking to what extent science can have the best of both worlds. Can we have Kuhnian normal science with its shared vocabulary, shared approaches to how questions should be asked, along with a Popperian openness to criticism? Might there be a way of setting up the incentive structure of science so that it works more effectively? Even people who are political conservatives, who have no sympathy with socialism in politics, have sometimes envisioned science as a kind of socialist paradise about intellectual matters, with everyone contributing to the common good. That's a more plausible picture about science, some have thought, than it is about economics. One could try to make that case by appealing, for instance, to a naturalistic investigation of the motives that draw people into science—but the case needs to be made, not assumed.

So far, we've asked questions about non-epistemic factors that, in some sense, are internal to science, how science is governed from within. But questions about the values and the social structure of science loom even larger when we turn our attention to science's role in society at large. For simplicity's sake, let's focus on publicly funded science. Privately funded science seems to have some legitimacy for serving the narrower interests by which it's funded, but it comes to figure in the public sector when a privately financed proposal gets put forward as some kind of scientific truth or is appealed to to guide public policy. In recent years (this is a major development in how universities are organized), we have seen a sharp rise in public/private partnerships where it's not clear whose interests the scientific research is supposed to be serving (sometimes that's a good thing; sometimes it's a bad thing, one might think).

It's sometimes claimed that science itself is morally neutral—that it's a pure kind of inquiry, and it's only technological applications of science that raise moral questions. This seems at least somewhat too simple. There seem to be some cases of scientific experiments, like those performed by Nazi doctors on victims in the Holocaust, where it's pretty clear that there's no gain remotely proportional, in scientific terms, to the suffering that is to be expected. It seems quite cavalier and dishonest to appeal to some notion of pure inquiry in such cases. That's a particularly grotesque example that doesn't falsify, perhaps, a more modest claim of scientific disinterestedness and objectivity.

But even in the less grotesque cases, we might not want to distinguish basic science too sharply from technology. To what extent do scientists have an obligation to reflect on the likely uses of their research? It's one thing to make an argument that it was morally appropriate to work on atomic weapons for the United States during World War II (that's a complicated issue about what Germany was up to, and saving lives that would otherwise be lost in various places—a difficult, complex matter).

That's quite different from trying to make the argument that the pursuit of knowledge itself was justified and that the moral consequences can be left entirely in the hands of those who will apply one's research. Maybe someone can make that case, but we generally hold people who play a crucial role in a questionable process to some standard of reflecting on whether it's appropriate to play the crucial role in that process. We expect people to at least ask (we're prepared to accept their answers) whether it's morally appropriate for them to be involved in the research they're involved in.

The case can be made from the other side that certain kinds of knowledge, anyway, are valuable for their own sake. It's kind of crazy to think that knowledge, as such, is always valuable for its own sake. Knowledge of the number of hairs on my head doesn't seem to be valuable at all. But arguably, some knowledge is valuable for its own sake, and in addition, lots of knowledge has unexpected applications, both for better and for worse. When scientists want huge sums of money for superconducting supercolliders and things like that, they often don't have a clear sense of any practical benefits—besides the knowledge that they think is itself scientifically valuable that's going to arise from the work. But nobody expected computers and the Internet to arise from some really quite abstract work in the logic of computation, either. So, there's a case to be made that the open-ended pursuit of knowledge tends to have good consequences.

If consequences of scientific work are generally unpredictable, the traditional view, anyway, shields scientists from moral responsibility; their job is just to get knowledge and to leave it to others to decide what is to be done with the knowledge. That's the traditional view.

There are lots of interesting test cases for views in this neighborhood. There's some decent evidence, for instance, about how data from the human genome project might well get used. It doesn't seem right to hold biologists responsible if people in 20 years are denied jobs or health insurance on the basis of a genetic profile. That's not the biologists' fault, but nor does it seem entirely irrelevant to raising questions about to what extent one wants to be involved in research that has a decent chance of being used that way.

There are also important issues about how scientists obtain their data. In our country, if people participate in a medical study you're running, it's generally accepted that you owe them the highest standard of medical care. To some extent, research is done in other countries for purposes of avoiding this expensive burden. So, one might worry, on the one hand, that we're using economically disadvantaged people as guinea pigs—taking advantage of already-huge inequalities in order to make our own medical research cheaper. On the other hand, there's a case to be made that we're giving these people better medical care than they would otherwise get if the study hadn't been conducted in their country. Philosophers of science generally leave these issues in the capable hands of ethicists. Philosophy of science, as such, does not have the resources to bring the right theories to bear on these questions.

Finally, we can note some difficulties about scientific decision-making. We might recall that Feyerabend thought science was becoming a threat to democracy. Why? Because non-scientists have to rely on scientists in order to ascertain the scientific significance of a proposal like the superconducting supercollider. But the scientists have an interest in, as it were, overselling the importance because they want to build the superconducting supercollider. Scientists don't seem similarly beholden to non-scientists to determine social significance. We don't have experts on what's socially important; we let each citizen have a full say in those matters, but scientists claim (we think rightly) more of a say about scientific significance.

So, how are we to balance these norms? Who should decide whether the superconducting supercollider gets built, and how are citizens supposed to weigh in on the social significance of such a project, given that they lack the scientific expertise that seems required for doing so?

The challenge is to set up a social structure that has the best chance of generating suitably informed judgments simultaneously of scientific and of social significance. This is a challenge that's really only recently been taken up in a kind of theoretically serious way about what kind of social structures might lead to the right kinds of decisions.

Next time we turn to a new topic that's really an old topic. Probability has glancingly figured in some of our earlier discussions, but in order to see what's currently central in philosophy of science, we need to make probability a theme rather than a side issue in our discussions.

Lecture Thirty
Probability

Scope:

Through much of Western intellectual history, "chance" was thought to represent the enemy, or at least the limitations, of reason. But notions of chance are now arguably inquiry's greatest ally. After a potted history of probability, we try to get clear about the basic mathematics of probability, and then we confront the philosophical issues that arise about the interpretation of probability statements. Such statements can be understood in terms of states in the world (for example, relative frequencies) or in terms of degrees of belief (for example, how likely you think it is that the Red Sox will win the World Series).

Outline

I. Probability has a fascinating history.
   A. The basic mathematical theory of probability did not really arise until around 1660.
      1. This seems quite shocking, given humanity's longstanding interest in gambling.
      2. Part of the reason seems to have been that chance did not seem like the sort of thing about which one could have a theory. The traditional Western conception of knowledge as modeled on geometry and as concerning that which must be the case probably played a role. Also, the Christian notion that everything that happens is a manifestation of God's will may have been a factor.
      3. The study of probability really got going when a nobleman and gambler asked Blaise Pascal to solve some problems about how gambling stakes could be divided up fairly.
      4. Probability caught on very quickly, if somewhat haphazardly, in business, law, and other applications.
   B. Arguably, probability is crucial to the modern conception of evidence.
      1. The term *probability* started off being associated with testimony. An opinion was probable if grounded in

reputable authorities. On that basis, it was not uncommon to hear it said that an opinion was probable but false, meaning that the authorities were wrong in that case.
2. Probability eventually morphed sufficiently to allow the necessitating "causes" of high sciences, such as physics and astronomy, to be assimilated to the mere "signs" of low sciences, such as medicine. The low sciences, lacking demonstrations, relied on testimony.
3. It was only in the Renaissance that the notion of diagnosis was distinguished from such notions as authority and testimony, on one hand, and from direct dissections and deductive proof, on the other. Probability becomes evidence when it becomes the testimony of the world, as it were. A symptom testifies to the presence of disease.
4. As the idea that physics, for example, could be demonstrative like geometry fades, we are left with an idea of evidence that derives from signs and symptoms. We have evidence of when one bit of the world indicates what another bit of the world is like.

C. In the 19th century, the spread of probabilistic and statistical thinking gradually undermined the assumption that the world was deterministic.
1. As governments kept better records of births, deaths, crimes, and so on, it emerged that general patterns could be predicted in a way that individual events could not.
2. As statistical laws became more useful, the assumption that they reflected underlying but virtually unknowable deterministic laws became increasingly irrelevant. The statistical laws started to seem the stuff of science, not a substitute for real science.
3. Important parts of statistical thinking migrated from such disciplines as sociology to such disciplines as physics, where, again, supposed deterministic explanations started to seem irrelevant.
4. With the arrival of quantum mechanics in the early 20th century, we encounter powerful arguments to the effect that our world is governed by statistical laws that are not backed by deterministic ones.

II. The mathematics of probability is uncontroversial. A somewhat casual sense of the mathematics will be adequate for our purposes.
  A. All probabilities are between 0 and 1.
  B. Any necessary truth gets assigned a probability of 1.
  C. If A and B are mutually exclusive, then the probability that one or the other will happen is equal to the sum of their individual probabilities.
    1. If there is a 30% chance that you will have the ranch dressing and a 40% chance that you will have the vinaigrette, then there is a 70% chance that you will have either the ranch or the vinaigrette (assuming, I hope correctly, that you'd never mix the two).
    2. Things get more complicated if the outcomes are not mutually exclusive. The chance that I will have the cake or the pie (given that I might have both) is the chance that I will have one plus the chance that I will have the other minus the chance that I will have both (essentially that is to avoid double-counting).
  D. Other rules for calculating probabilities can be built (roughly) from these.
III. Controversy arises in the interpretation of the mathematics. We will consider three major interpretations.
  A. *Frequency theories* place probabilities "out there" in the world. This is the most commonly used concept of probability in statistical contexts. The frequency theory identifies probabilities with certain relative frequencies.
    1. Probabilities could be construed as *actual* relative frequencies. The probability of getting lung cancer if you smoke is the ratio of smokers with lung cancer to the total number of smokers. This approach is clear and links probabilities tightly to the evidence for them.
    2. This approach faces issues about how to place objects in scientifically salient populations. The probability that I will get lung cancer is either 1 or 0. And I have one probability of getting lung cancer as a nonsmoker, another as a 40-year-old male, another as a coffee addict, and so on.

3. A more serious problem occurs because this account is "too empiricist." It links a scientific result too closely to experience. A coin that has been tossed an odd number of times cannot, on this view, have a probability of .5 of coming up heads. In addition, a coin that has been tossed once and landed on heads has, on this view, a probability of 1 of landing on heads. Such single-case probabilities are a real problem for many conceptions of probability.
4. One might go with *hypothetical* limit frequencies: The probability of rolling a seven using two standard dice is the relative frequency that would be found if the dice were rolled forever. We saw an idea like this in the pragmatic vindication of induction.
5. This version might not be empiricist enough. The empiricist will want to know how our experience in the actual world tells us about worlds in which, for example, dice are rolled forever without wearing out.

B. *Logical theories* treat probabilities as statements of evidential relationships. They can be interpreted as the judgments of an ideal agent or as relations in logical space. The idea here is that probability gives a logic of partial belief or inconclusive evidence modeled on what deductive logic provides for full belief or conclusive evidence.
1. Just as our full beliefs should not contradict one another, our partial beliefs should cohere with one another. Having coherent beliefs is not sufficient for getting the world right, but having incoherent beliefs is sufficient for having gotten part of it wrong.
2. Probabilistic coherence is a matter of how well an agent's partial beliefs hang together. If your evidence assigns a probability of .8 to *p*, then it had better assign a probability of .2 to *not-p*.
3. Logical theories of probability impose conditions beyond mere coherence. In particular, they impose the *principle of indifference*. If your evidence does not give you a reason to prefer one outcome to another, you should regard them as equally probable.
4. The mathematics of probability does not require this principle, and it turns out to be very troublesome. There are many possible ways of distributing indifference, and

it's hard to see that rationality requires favoring one of these ways.
   C. *Subjective theories* treat probabilities as degrees of belief of actual agents—they directly concern the believing agent rather than the world, but they are subject to objective although rather minimal criteria of rationality.
      1. A degree of belief is measured by one's notion of a fair bet. The odds at which you think that it would be reasonable to bet that a Democrat will win the next presidential election tell you the extent to which you believe that a Democrat will win.
      2. Because this approach does not explicate probabilities in terms of frequencies or a principle of indifference, it relies only on the notion of probabilistic coherence to make probability assignments "correct."
      3. For this reason, this model as so far described seems to allow any old probabilistically coherent set of beliefs to be perfectly rational. Paranoid delusions tend to be strikingly coherent yet seem to be rationally criticizable. We will address this problem in the next lecture.

Essential Reading:

Curd and Cover, "Bayes for Beginners," in Curd and Cover, *Philosophy of Science: The Central Issues*, pp. 627–638.

Hacking, *The Emergence of Probability*.

Supplementary Reading:

Hacking, *The Taming of Chance*.

Questions to Consider:

1. Geometry served as a paradigm of knowledge for centuries. What paradigms of knowledge operate in our culture at present? Do any of them reflect the shift discussed in this lecture to the idea that we can have knowledge of contingent matters?

2. If you knew that an urn consisted of red and green balls (but knew nothing else about it), would it be irrational to let the fact that you like red better than green affect your probability judgments? What kind of mistake, if any, would you be making

if you assigned a probability of .9 to drawing a red ball and .1 to drawing a green one?

Lecture Thirty—Transcript
Probability

Probability and some closely allied notions have cropped up a couple of times in this course, and, to a certain extent, we've treated them as distractions from the main business at hand. This is a situation for which I am fixing to apologize. We've already seen some indications that probability is more fundamental than this usual way of treating it as a kind of complicating factor would indicate.

So, in our discussion of induction and evidence, we saw that probability provides the way to understand evidence statements in general. "All copper conducts electricity" and "No particle travels faster than light" are limiting cases, special cases, in which the frequency of a trait in a sample is 100 percent (in the case of copper conducting electricity) or 0 percent (in the case of particles that travel faster than light). A lot of important stuff happens in between those extremes: Some smokers don't get cancer, and some non-smokers do. So, for lots of the work science is called upon to do, we need to deal with frequencies that are in between these extremes. What, exactly, probability statements have to do with these frequencies is an issue that will occupy us later in this lecture.

In our discussion of scientific explanation, we also saw that explanation works surprisingly differently when statistical laws, rather than deterministic ones, are in the picture. In Karl Hempel's model, anyway, a scientific explanation has to get relativized to background information when statistical laws are in place, in a way that deterministic laws don't require such relativization. Why? Because new information can render a previously strong probabilistic explanation extremely weak. If we learn that our patient is allergic to penicillin, our prediction that he will recover from his infection goes from very probable to very improbable with the addition of new information.

These complications involving probabilistic explanation are especially unfortunate given that quantum mechanics gives us some reason to believe that the most fundamental laws governing the entire universe are probabilistic rather than deterministic. According to quantum mechanics, there is no cause that explains why one uranium atom decays and an identical one doesn't; there are just brute probabilistic laws that a certain percentage decay within a certain time. If you insist on some explanation, empiricists will tell you you're pushing the demand for explanation too far. There is no deeper explanation then a brute probabilistic law.

So, probability and statistics (we won't need to carefully distinguish these notions for our purposes) more or less represent the normal case. The deterministic case is the falling off from the probabilistic norm. We tend to focus on non-probabilistic cases because, in many respects, they're simpler, but doing so can distort important features of the general case of scientific reasoning. So, feminists complain sometimes that male doctors used to treat distinctively female organs as distractions from what a normal body is like. That's a bad way to practice medicine and, similarly, treating probability statements as a falling off from a norm of determinism gets philosophy of science, in some important respects, backwards. It's perhaps forgivable since philosophy of science is hard and we wanted to avail ourselves of every simplifying opportunity that we could find. But, nevertheless, we've got to confront the fact that general cases are probabilistic rather than deterministic.

So, in this lecture and the next two, we start to rectify the situation. We'll focus mainly on issues of confirmation and evidence, and we'll see that probability makes an enormous difference there. That starts next time. This time, we start to get clear about the basic notions in the field of probability. As the English philosopher Bishop Joseph Butler famously said, "Probability is the very guide to life."

Probability has a fascinating history, and the idea that it could be a guide of life was rather slow in coming. The basic mathematical theory of probability did not really arise until around 1660. This doesn't sound shocking by itself, but it is shocking when you realize that gambling had been going strong for thousands and thousands of years by this time.

Ian Hacking—perhaps the foremost philosopher concerned with the history of probability—cracks that a gambler with basic knowledge of the mathematics of probability could have won all of Gaul within a week. So, given how useful the basic mathematics of probability would have been, it's perhaps striking that it didn't develop until so much later in Western history.

Absences are usually rather difficult to explain, and—as far as I know, anyway—there isn't a generally accepted explanation of why something as useful as a basic understanding of probability took so long to emerge.

We are venturing a bit into intellectual history here, and I have to rely on standard sources because I'm not an intellectual historian; I have no special expertise. It's widely believed, anyway, that part of the reason that no theory of probability arose goes pretty deeply into Western culture. The idea is that chance didn't seem like the sort of thing that was amenable to being understood. You're not going to have a theory of something that seems beyond comprehension.

Joseph Bertrand, an important mathematical theorist of probability, was able to write as late as 1888: "How dare we speak of the laws of chance? Is not chance the antithesis of all law?" So, if chance is a name for ignorance, for limits to knowledge, that suggests—though it by no means establishes—that it's not itself the sort of thing that can be known. That's going to be a discouraging factor when you try to sort-of get your head around probability.

As we briefly noted back at the beginning of the course, for much of Western intellectual history, knowledge was of that which could be demonstrated, that which had to be the case. The model of real knowledge is geometry. If that's your picture of knowledge, demonstrations of things that have to be the case, that flow from some kind of self-evident principle, then chance will, by its very nature, seem to be unknowable.

In addition, some scholars have speculated that the Christian notion that everything that happens is a manifestation of God's will, and so is determined, though in ways that we can't begin to understand, might also have discouraged systematic reflection on chance phenomena.

Whatever the explanation, sophisticated probabilistic reasoning seems to have caught on much more quickly in India than in Europe. As Hacking emphasizes, however, we may be better off asking how we came to have our concepts of probability rather than why these concepts didn't show up sooner. His suggestion is that it is by no means inevitable that the phenomena around the notion of probability would get divided up as they did—that it's kind of a gerrymandered notion (this will become at least slightly clearer as we proceed).

The study of probability really got going when a nobleman and gambler asked the French mathematician and philosopher, Blaise Pascal, to solve some problems about how you divide up a gambling stake fairly when the game is called off and everybody goes home (who gets how much of the pot and why?).

Once the mathematics started to develop, probability caught on very quickly, if somewhat haphazardly, in business (especially in businesses like insurance). But it really was haphazard—you found governments selling annuities for 100 years after the rise of probability theory that didn't take the age of the people buying essentially life insurance into account. They were trying to use the mathematics of probability, but didn't seem to have a deep understanding of how it was supposed to work.

Our concern, however, is with probability's role in science. Arguably, probability is central to the modern conception of evidence. The word "probability" starts off being associated with testimony. An opinion was probable if it was grounded in reputable authorities. Given that meaning of the term "probability," it was not uncommon to hear it said that an opinion was probable but false. What could that mean? It meant that the authorities were wrong in this case—so the opinion was supported, but not in a way that made it likely to be true. That's not what "probability" meant. It meant the authorities, as it were, stood with the opinion.

Probability eventually morphs enough to allow the "causes" of high sciences like physics and astronomy—which reasonably aspire to this geometrical model—to get assimilated to the "signs" of low sciences like medicine, which could not approach real demonstrative knowledge.

"High" sciences, as we've seen, aspired to be systematic, while "low" sciences—which included geology and alchemy, as well as medicine—had to settle for much less. They could deal only in opinion, which was a radically different kind of thing than knowledge. So, they had to rely almost entirely on testimony, at least in their original incarnations.

It's only in the Renaissance that the notion of diagnosis—of a thing serving as an indication of another thing—came to be clearly distinguished from notions like authority and testimony on the one hand, versus direct dissections in medicine and deductive proof on the other. It occupied this weird nether region between real evidence and relying on somebody else's evidence.

Probability becomes what we now regard as evidence when it becomes the world's testimony, as it were. A symptom testifies, in an all but literal sense, to the presence of the disease. It's a report, but it's not a report from another person; it's a report from the world. It's through that morphing of the concept of probability—from testimony to the world's testimony—that we get what we now regard as our notion of evidence.

As the idea that physics, for instance, could be demonstrative like geometry starts to fade (we've looked at that a little bit in this course), we end up with an idea that anything worth calling evidence derives from signs and symptoms—that it's one bit of the world indicating inconclusively but importantly what another bit of the world is like. So, the notion of evidence derives from second-class sciences taking over, and the first-class sciences modeling themselves on the second-class sciences.

In the 19th century, the spread of probabilistic and statistical thinking gradually undermined the assumption that the world was deterministic.

The history of this is rather surprising as well. As governments kept better records of births, deaths, crimes, and suicides, it emerged that the general patterns could be predicted, had regularities, were lawful, in a way that individual events could not be predicted, did not seem to be lawful. We could know approximately how many people would die violently in France in a given year, even though we couldn't predict which people they were or how exactly it would happen.

The assumption all along was that we settled for statistical laws because we couldn't understand the presumably fully determinate processes leading to each individual homicide. But as the statistical laws became increasingly useful, the assumption that they reflected underlying but more or less unknowable deterministic laws became increasingly irrelevant. All of the predictive juice is contained in the statistical laws. It's just a kind of metaphysical assumption that there's determinism backing those laws. The statistical laws started to seem to be the stuff of science, not a substitute for real science.

So, in the social or human sciences, the idea of a human nature shared by all people starts to morph into the idea of "the average person" or "the normal person" with predictable deviations from the statistical norm. The normal person was sometimes thought to be the best specimen of the species and was sometimes thought to be a distinctively mediocre specimen of the species. But it was thought, in either case, that the numbers could—if we could just compile them in the right way—simultaneously settle "is" questions about society and "ought" questions about society. Because the notion of a norm is descriptive and normative—it's the way things, in some sense, should be.

It was something of a two-say street, but it is fair to say that important parts of statistical thinking migrated from disciplines like sociology to disciplines like physics, and then deterministic explanations started to seem increasingly irrelevant in physics. So, again, we have innovation flowing, in this case, from a supposedly "soft" science to a supposedly "hard" science. Again, it flowed in both directions, but innovations don't always track sort-of prestige and scientific success.

As we've seen, at least glancingly, with the arrival of quantum mechanics early in the 20th century, we start to encounter powerful scientific arguments to the effect that our world is governed by statistical laws that are not—and maybe cannot be—backed by deterministic laws. As we'll see in a few lectures, you don't actually have to get as far as quantum mechanics to start to see a more or less irreducible role for probability in physics. The reduction of thermodynamic phenomena (things like heat and entropy) to the motion of molecules involves treating this part of physics as inescapably, if not perhaps irreducibly, probabilistic (we'll talk about that in a couple of lectures).

Notice that we did all of this potted intellectual history without clarifying the notion of probability. That was intentional, because the notion of probability eluded clarification throughout most of its history—and, to some extent, it still does so today. The notion of probability has received a ton of attention in mathematics and philosophy in the 20th century, but there is no agreement about the basic interpretation of probability. There's a range of basic interpretations, but which one is most useful for scientific purposes is a very vexed question.

There is, however, a common mathematical core that all probability theorists agree about. We need to characterize this, though we'll do it somewhat loosely; mathematical rigor is beyond our needs in this lecture.

All probabilities are between 0 and 1. That's a definitional claim; you can think of it as probabilities being between 0 percent and 100 percent, if you prefer. That's, of course, equivalent.

Any necessary truth gets assigned a probability of 1. You might agree with Quine, back in Lecture Eight, that no statement should be considered necessary or unrevisable. Even the laws of logic, Quine thought, might get revised in the course of inquiry. It's okay if you think that. The point is just that if any statement were to express a necessary truth (if you thought the statement "There are no square circles" is a necessary truth), then it should get a probability of 1. That's built into the notion of probability.

In addition, we need a rule for adding probabilities, and everything else will more or less flow from these axioms.

If A and B are mutually exclusive, then the probability that one or the other will happen is equal to the sum of their individual probabilities. If there is a 30 percent chance that you will order the Ranch dressing and a 40 percent chance that you'll order the vinaigrette, then there's a 70 percent chance that you'll have either the Ranch or the vinaigrette (I'm helping myself to the assumption that you would not be so disgusting as to put both of them on your salad at once).

It gets more complicated if the probabilities are not mutually exclusive. The chance that I'll have the cake or the pie, given that I might have both, is more complicated. It's the chance that I'll have the cake plus the chance that I'll have the pie, minus the chance that I'll have the cake and the pie (because we essentially have to avoid double counting).

The other rules for calculating probabilities can be built out of these axioms. You actually only need the rule for adding mutually exclusive outcomes; you can build everything else that you need mathematically out of those three principles.

But so far, this is just a mathematical formalism (with some kind-of silly examples thrown in). It just gives us the laws anything will have to obey in order to count as a probability function. But lots of mathematical functions that aren't intuitively probability functions obey this. Length can be made to obey a kind of probability function. We only refer to this kind of mathematical entity as a probability function when we give it an interpretation in certain respects.

So, the issue is what gets modeled by the mathematics. What are probability statements about? That's a philosophical question about probability. We will consider three major interpretations—there are a few others, but these will suffice for what we need.

The first are *frequency theories*. These place probabilities "out there" in the world. A probability statement is about facts, as it were. This is the most commonly used conception of probability, especially in statistical contexts.

As gamblers and actuaries (or at least good gamblers and actuaries) know, probabilities had better have something to do with frequencies. If the probability of drawing a straight flush in a certain poker game had nothing to do with how frequently it happened, we would have no use for the statement about the probability of drawing a straight flush. So, the frequency theorist makes a simple identification. We know probabilities must have something to do with relative frequencies. The frequency theorist identifies probabilities with relative frequencies. There's more than one way to do that, however, so there's more than one frequency interpretation of probability.

Most straightforwardly, probabilities could be construed as actual relative frequencies. This very directly makes probability statements factual statements. The probability of getting lung cancer if you smoke is the ratio of smokers with lung cancer to total smokers. That's oversimplified, of course. In a real scientific case, we'd need to know how much one smokes, for how long one has smoked, and we would have to add a few other things, but the point is that it's an actual relative frequency of a trait in a population. This has certain very impressive virtues: It's clear, and it has a clear connection to empirical evidence. We know what we're saying when we make a probability statement, and we know what backs the statement.

We'll face some issues about why such probability statements are scientifically interesting. As we saw when we looked at statistical explanation, we have to place a given individual in a scientifically relevant population in order to make the relative frequencies scientifically interesting.

I have one probability of getting cancer as a non-smoker, a different one as a 40-year-old male, perhaps a different one as a coffee addict. If you just focus on me, the probability that I will get a given kind of cancer is either 1 or 0, if we assume that the laws governing that kind of phenomenon are deterministic.

So, what do we mean when we talk about my probability for getting a certain kind of cancer? Generally, unless we specify a narrower context, what we mean is we want to use all of the information that's statistically relevant. So, we don't need to know the rate of cancer for people who have my exact number of freckles, for instance. The point is, that despite the actual relative frequency interpretation, making probability statements about things "out there" in the world, their usefulness nevertheless depends on a state of information. There is no such thing (other than perhaps the 1 or 0 probability, which is *my* probability of getting cancer); the only meaningful statement here is how I'm described or categorized, what reference class I am placed in. That's what gives meaning to the actual relative frequency.

As stated, this actual relative frequency interpretation of probability faces a very serious problem. Speaking kind of loosely, but drawing on the background we've developed in this course, it's "too empiricist." Like an operational definition or a reduction sentence, back in our positivistic days, it links a scientific result too closely to experience. A coin that has been tossed an odd number of times cannot—on an actual relative frequency view—have a probability of .5 of coming up heads, because it's been tossed an odd number of times. Similarly, a coin that's been tossed once has a probability either of 1 or 0 of landing on heads. We don't think that captures the facts about probability. It reduces the probability statement too thoroughly to the evidence for the probability statement.

We should note that single-case probabilities (talking about the probability of an event that is not repeatable) present problems to every interpretation of probability. But nevertheless, it's a particularly grotesque problem for the actual relative frequency view, since that seems seriously to misstate what we want to say about the probability of the coin landing on heads.

So, one might go with a *hypothetical limit frequency*. We might say, for instance, that the probability of rolling a seven using two standard six-sided dice is the relative frequency that would be found if the dice were rolled forever (assuming that there's an answer). We saw when we looked at the pragmatic vindication of induction that induction can be shown to work if there's a long-term limit frequency for a trait in the population. So, assuming that there's an answer, that answer is your probability of getting a seven when you roll two dice.

But this version doesn't seem to be empiricist enough. We've seen that empiricists raise serious worries about what we called *counterfactual conditionals*, or *contrary-to-fact conditionals*. These are "if, then" statements about how the world would be if things were different than they, in fact, are. But the dice, and maybe the whole world, would have to be pretty different if the dice were to be rolled forever. For instance, people would have to have much longer attention spans than they do if people are going to role these dice forever and write down what happens. The sun would have to last indefinitely if these dice were to be rolled forever. So, for this interpretation to work, we need to find a way to give clear meaning to the counterfactual conditional, what it would take for these dice to be rolled forever. We need some answer to that question.

So, it's obvious and important that probabilities have something to do with frequencies. But to identify them with actual relative frequencies seems too reductively empiricist. And to get away from the actual relative frequencies builds in idealizing conditions that get pretty far away from the evidence. That's a problem.

Logical theories treat probability statements as statements about evidential relationships. This is a pretty different approach than the frequency approach. On this interpretation, probability statements can be thought of as the judgments of an ideal agent or as evidential relationships, as it were, in logical space.

The idea is that probability gives a kind of logic of partial belief or inconclusive evidence modeled on—but generalizing from—deductive logic. Deductive logic tells us how different statements bear on one another when the statements are regarded as having probabilities of 1 or 0, when they're either true or false. So, we generalize deductive logic to cover all of the cases in which the evidence neither conclusively establishes nor conclusively refutes the statement.

This is the inductive logic that the logical positivists, especially Rudolf Carnap, pursued with such energy. So, we know that it didn't work (we've gotten to that point in our course).

The first component of this approach is probabilistic coherence. Just as our full beliefs shouldn't contradict one another, our partial beliefs should cohere with one another. Having coherent beliefs is not sufficient for getting the world right, but having incoherent beliefs is sufficient for having gotten at least part of the world wrong. So, we'd like to avoid having either deterministically or probabilistically incoherent beliefs.

Contradictions are harder to come by with partial belief. My belief that the Red Sox might win doesn't contradict my belief that the Yankees might win, as it would if I took out the "might" (assuming the teams are playing each other). So, the notion of a contradiction doesn't carry over in quite the same way to the probabilistic case.

Probabilistic coherence is, instead, a matter of how well an agent's partial beliefs hang together. If your evidence assigns a probability of .8 to a statement, then it had better assign a probability of .2 to the negation of the statement. Similarly, you'd better not think it's 60 percent likely that the Red Sox will win and 60 percent likely that the Yankees will win when they're playing each other. The probabilities for mutually exclusive outcomes should sum to 1. There's something incoherent (in a way we'll explore more deeply next time) about having probability assignments that don't meet this condition of coherence.

But logical theories of probability impose conditions of rational belief that go beyond mere coherence. In particular, they impose what's called the *principle of indifference*. If your evidence does not give you a reason to prefer one outcome to another, you should regard them as equally probable. So, if you know absolutely nothing about baseball, you should assign a probability of .5 to each team winning. That's the probability assignment that properly reflects your evidential situation, and so it's the rational probability estimate to adopt. Doing anything else would reflect a kind of irrational bias.

Note that no principle of indifference was built into the mathematics of probability; this is an independent condition of rationality that logical theorists of probability impose. And it turns out to be very troublesome—as Joseph Bertrand, from whom we heard earlier in this lecture, showed.

Let's say that I have a vase, and I don't know how much water is about to be poured into it. So, I'm supposed to distribute my probability assignments in a way that shows my ignorance. But am I supposed to be indifferent between, because I'm ignorant of, various heights of water that could be poured into the vase or, say, various volumes of water that could be poured into the vase? Unless the vase has a linear shape, those are different ways of distributing these prior probabilities, different ways of reflecting my ignorance. The idea that one of them is rational and the other is irrational seems arbitrary and ungrounded.

If I know that the pizza I'm about to order will be within a certain size range, should I distribute my ignorance over possible pizza diameters or over possible pizza areas? In the absence of any evidence, is one of these supposed to be the rational way to reflect my ignorance and the other not?

This is very like the lesson we learned from Nelson Goodman's new Riddle of Induction. Goodman's Riddle showed that there are too many regularities out in the world; there are too many properties that could count as the color of emeralds (they could be "green"; they could be "grue"; they could be "gred"). Bertrand shows that there are too many outcomes we might be indifferent between when we're in a state of ignorance. We could be indifferent between diameters, or areas, or volumes. In neither case—Goodman's or Bertrand's—will reason or language tell us which properties or which possibilities matter. To say it's a rule of rationality that you use a certain language to divide up the space of possibilities seems arbitrary and undermotivated.

This suggests something like the logical approach to probability that drops the principle of indifference. These are often called *subjective theories of probability*. They treat probabilities as degrees of belief, but of something more like actual agents than ideal agents.

We saw that the logical theory of probability is easiest to model as degrees of partial belief of a perfectly rational or ideal agent because the evidential relationships, the kind of logic, gets messy. The subjective theory just says we're modeling the beliefs of actual agents. Probability statements concern the believing agent, rather than facts out in the world, as the frequency theory has it. These probabilities are subject to objective—but rather minimal—criteria of rationality.

So, what do we mean by a "degree of belief" or a "degree of partial belief"? The standard way of measuring it (and this is a kind of operational definition) is by one's notion of a fair bet. The odds at which you think (we're talking about your degrees of belief here) that it would be reasonable for someone to bet that a Democrat will win the next presidential election tells you the extent to which you believe that a Democrat will win. The more unlikely you think it is, the higher a payoff you would demand before taking the bet.

Notice that even here, we're implicitly linking degree of belief, this fact that's somehow "in here," with some notion "out there" of frequency and repetition. What makes the betting behavior a good measure of belief is the idea that, by your lights, you would expect to break even if you took a large number of such bets. You'd lose some of the time, but win other times, and if the payoff is enough for an unlikely bet, you'd break even. That's what makes it a fair bet; that's what makes it a manifestation of your degree of belief. So, we're constantly tempted to use frequencies, to use repetition, as tests for correctness of degree of belief.

But since we've dropped the principle of indifference, this approach does not explain probabilities in terms of frequencies or rules of rationality. It relies only on the notion of probabilistic coherence (at a time and across time [this will be central in our discussion next time]). That's all it takes to make probability assignments as "correct" as they get.

The problem here is, it seems way too easy to have probabilistically coherent beliefs. Paranoid delusions are strikingly coherent ("Everybody's out to get me," and I will fit everything into that framework). A paranoid delusion is probabilistically coherent. I assign a high probability to "They're out to get me" and a low probability to "They're not out to get me." But it seems like a fatally flawed web of belief.

So, our task next time is to see how an effective approach to scientific reasoning can be built out of such modest resources. Requiring only that beliefs be probabilistically coherent at a time and across time gives us an extremely influential approach to scientific problems of confirmation and evidence.

Lecture Thirty-One
Bayesianism

Scope:

Bayesian conceptions of probabilistic reasoning have exploded onto the philosophical and scientific scene in recent decades. Such accounts combine a subjectivist interpretation of probability statements with the demand that rational agents update their degrees of belief in accordance with *Bayes's Theorem* (which is itself an uncontroversial mathematical result). Bayesianism is a remarkable program that promises to combine the positivists' demand for rules governing rational theory choice with a Kuhnian role for values and subjectivity. After explaining and motivating the basics of Bayesianism, we examine its approach to scientific theory choice and to the raven paradox and the new riddle of induction.

Outline

I. Starting from very modest resources, the *Bayesian* approach to probability has rejuvenated philosophical thinking about confirmation and evidence.

   A. Bayesianism begins with a subjective interpretation of probability statements: They characterize personal degrees of belief. These degrees of belief can be more or less measured by betting behavior; the more unlikely you think a statement is, the higher the payoff you would insist on for a bet on the truth of the statement.

   B. Your degrees of belief need not align with any particular relative frequencies, and they need not obey any principle of indifference. Bayesianism requires little more than probabilistic coherence of beliefs.

   C. The *Dutch book argument* is designed to show the importance of probabilistic coherence. To say that a Dutch book can be made against you is to say that, if you put your degrees of belief into practice, you could be turned into a money pump.

   1. If I assign a .6 probability to the proposition that it will rain today and a .6 probability to the proposition that it

will not rain today, I do not straightforwardly contradict myself.
2. The problem emerges when I realize that I should be willing to pay $6 for a bet that pays $10 if it rains, and I should be willing to pay $6 for a bet that pays $10 if it does not rain.
3. At the end of the day, whether it rains or not, I will have spent $12 and gotten back only $10. It seems like a failing of rationality if acting on my beliefs would cause me to lose money no matter how the world goes.
4. It can be shown that if your degrees of belief obey the probability calculus, no Dutch book can be made against you.

II. But pretty loony webs of belief can still be probabilistically coherent. Bayesianism becomes a serious scientific theory of scientific rationality by developing a theory of how one should handle evidence. The first component of this theory is a notion of confirmation as raising the probability of a hypothesis.

A. Bayesians think that the notion of confirmation is inherently quantitative. We cannot ask whether a piece of evidence, E, confirms a hypothesis, H, unless we know how probable H started out being—we have to have a *prior probability* for H. E confirms H just in case E raises the prior probability of H. This means that the probability of H given E is higher than the probability of H had been: $P(H/E) > P(H)$. E disconfirms H if $P(H/E) < P(H)$.

B. All this is done within the subjectivist or personal interpretation of probability. A big cloud on an otherwise clear horizon counts as evidence of rain for me, just in case my subjective probability that it will rain, given the new information that there is a big cloud on the horizon, is higher than my prior probability that it would rain.

C. In saying this, we have made tacit use of the notion of *conditional probability*: the probability of the hypothesis conditional on or given the evidence.
1. The conditional probability of H given E is the probability of (H&E) divided by the probability of E (provided that E has a nonzero probability). (H&E) is the intersection, the overlap, of cloudy days and rainy days.

The definition says that the higher the percentage of cloudy days that are rainy, the higher the conditional probability of H given E.

   2. If I were already convinced that it would rain (because of a weather report, for instance), then this high conditional probability of rain depending on clouds would not change my prior belief and, thus, would not be evidence. But, if I had been relatively neutral, it might significantly confirm the rain hypothesis for me.

   D. The idea that whatever raises the probability of H confirms H is not without its problems. My seeing Robert De Niro on the street might raise the probability that he and I will make a movie together, but it hardly seems to count as evidence that we'll make that movie. However, we'll assume that such problems can be solved.

III. The second idea crucial to Bayesians is that beliefs should be updated in accordance with *Bayes's Theorem*.

   A. The theorem itself is a straightforward consequence of the definition of conditional probability. Non-Bayesians accept the truth of the theorem but don't put it to the use that Bayesians do.

   B. The classic statement of the theorem is:
   $$P(H/E) = \frac{P(E/H) \times P(H)}{P(E)}.$$

   C. The left side of the statement is the conditional probability of the hypothesis given the evidence. It can have two different readings, depending on whether the evidence is "in" yet or not.

   1. If the evidence is not in, then $P(H/E)$ is the prior conditional probability of H given E. If I were a physicist in 1915, I might have assigned a low probability to Einstein's hypothesis of general relativity, but I also might have thought to myself, "If it turns out that light rays are bent by the Sun, I assign a quite high probability to Einstein's hypothesis."

   2. If the evidence is in, then $P(H/E)$ represents the posterior probability of the hypothesis. It is the probability I now

assign to Einstein's hypothesis, once I have gotten news that light rays are bent.
- **D.** We now unpack the right side of the statement.
  1. $P(E/H)$ measures how unsurprising the evidence is given the hypothesis. Given Einstein's hypothesis of general relativity, the probability that light rays are bent by the Sun's gravitational field is quite high.
  2. $P(H)$ is just the prior probability of the hypothesis.
  3. The posterior probability (that is, the left side of the equation) is directly proportional to the prior probability of the hypothesis and directly proportional to the extent to which the hypothesis makes evidence unsurprising.
  4. The prior probability of the evidence is the denominator of the fraction, reflecting the fact that, all other things being equal, unexpected evidence raises posterior probabilities a lot more than expected evidence does. Apart from Einstein's theory, the probability of light being bent by the Sun was quite low. It is because Einstein's prediction is so unexpected, except in light of Einstein's theory, that the evidence had so much power to confirm the theory.
  5. Thus, the more unexpected a given bit of evidence is apart from a given hypothesis and the more expected it is according to the hypothesis, the more confirmation the evidence confers on the hypothesis.
- **E.** The controversial part arises when the Bayesian proposes as a rule of rationality that, once the evidence comes in, the agent's posterior probability for H given E should equal the agent's prior conditional probability for H given E.
  1. This sounds uncontroversial; as we saw, there were just two interpretations of the left side of the equation. But the mathematics by itself will not get you this result.
  2. Once the evidence comes in, I could maintain probabilistic coherence by altering some of my other subjective probabilities, namely, some of the numbers on the right side. I could decide that the evidence was not that surprising after all, for instance, thereby making my posterior probability different from my prior conditional probability. Why must today's priors be tomorrow's posteriors?

F. The Bayesian appeals to a diachronic (across time) Dutch book argument to support this requirement. If you use any rule other than Bayesian conditionalization to update your beliefs, then a bookie who knows your method can use it against you by offering you a series of bets, some of which depend on your future degrees of belief.

IV. Bayesianism has helped rekindle interest in issues about evidence and justification. The Bayesian approach allows for impressive subjectivity (there are very few constraints on prior probabilities other than coherence with other degrees of belief) and impressive objectivity (there is one correct way of updating one's beliefs in the face of new evidence).

A. Bayesians argue that initial subjectivity disappears when enough good evidence comes in. This is called the *washing out of prior probabilities*. It can be established that no matter how great the disagreement is between two people, there is some amount of evidence that will bring their posterior probabilities as close together as you would like. That is impressive, but it is subject to some significant limitations.

1. If one person assigns a prior probability of 0 to a hypothesis, no evidence will ever increase that probability.

2. There is no assurance that convergence will happen in a reasonable amount of time.

3. The washing-out results require that the agents agree about the probabilities of all the various pieces of evidence given the hypothesis in question. This seems problematic.

B. Bayesianism's attractiveness as a theory of scientific inference can be appreciated by revisiting Goodman's new riddle of induction and Hempel's raven paradox.

1. The Bayesian will say that there is nothing the matter with either of the new riddle's inductive arguments. It is fine to infer from the greenness of emeralds to their continued greenness or from their "grueness" to their continued "grueness." Whichever hypothesis you think more probable going in will remain more probable going out.

2. Bayesians can handle the raven paradox equally straightforwardly. The greater the ratio of P(E/H) to P(E), the greater the power of evidence to confirm H. This turns out to be the source of the difference in the confirming power of white shirts and black ravens to confirm "All ravens are black." The probability that the next raven I see will be black given that all ravens are black is 1. The probability that the next shirt I see will be white given that all ravens are black is much lower. It is pretty much just my prior probability that the next shirt I see will be white.

Essential Reading:

Salmon, "Bayes's Theorem and the History of Science," in Balashov and Rosenberg, *Philosophy of Science: Contemporary Readings*, pp. 385–404.

Godfrey-Smith, *Theory and Reality: An Introduction to the Philosophy of Science*, chapter 14.

Supplementary Reading:

Salmon, "Rationality and Objectivity in Science or Tom Kuhn meets Tom Bayes," in Curd and Cover, *Philosophy of Science: The Central Issues*, pp. 551–583.

Questions to Consider:

1. Utter fictions can be quite coherent. Does the Bayesian need an argument that a set of beliefs that is probabilistically coherent (both at a time and across time) is likely to be true? Does the Bayesian have resources to provide such an argument?

2. Does the Bayesian solution to Goodman's new riddle of induction seem satisfactory to you? The solution works if you have the right prior probabilities, but it doesn't claim that you *should* have those prior probabilities.

Lecture Thirty-One—Transcript
Bayesianism

I feel obliged to begin this lecture with a kind of modest warning because we are fixing to commit some mathematics in here. I believe it will be intelligible for audio customers without too much difficulty, and I would encourage people not to pay exceptionally close attention to the mathematics if they're finding it difficult or distracting. The points that are needed will get stated non-mathematically either right before or right after the mathematics. But I do want to get, for the record, the formulations that we're officially talking about before you. Nevertheless, I would like to encourage anybody driving in big-city traffic to just let the math go if it's requiring too much concentration; I don't want to be responsible for any accidents.

I'd also like to assure people that I do not perpetrate mathematics lightly. I think this is going to be a very useful avenue into some problems of real human heft and significance. I'll try intermittently in this lecture, and quite substantially at the end of the next lecture, to indicate why I think the mathematics is an avenue into something that is, by no means, dry and technical, but really quite immediately important.

Starting from very modest resources, the *Bayesian* approach to probability has rejuvenated, in recent decades, philosophical thinking about confirmation and evidence. I'll explain why it's called *Bayesianism* in a little while.

Bayesianism begins with what we called last time the *subjective interpretation of probability statements*. A probability statement reports or characterizes personal degrees of belief. These degrees of belief are more or less measured by betting behavior; the more unlikely you think a statement is, the higher the payoff on which you'd insist in order to accept a bet on the truth of the statement.

Orthodox Bayesians don't require your beliefs to line up with any observed frequencies, any supposed hypothetical relative frequencies, and they don't impose a kind of logical principle of indifference, as the logical interpretation of probability did last time.

So, subject to a few additional constraints, which we'll impose shortly, your degrees of belief for an orthodox Bayesian are your own business. If, for whatever reason, you decide that it's unlikely that copper conducts electricity, Bayesianism, so far anyway, has no gripe with that estimate, as long as it coheres with your other beliefs—so you'd better not then assign a high probability to metals conducting electricity, unless you then assign a low probability to copper being a metal. However you want to maintain a coherent belief set is okay with the Bayesians—coherence is the heart of rationality.

Deductive incoherence means that your beliefs contradict one another. Probabilistic incoherence is a bit trickier. The idea that this kind of coherence is a requirement of rationality is generally established by what is called the *Dutch Book Argument*. The origin of this term is shrouded in mystery. I, anyway, do not intend any offense to our Netherlandish friends, but it may well have been a slur back in the day.

To say that a Dutch Book can be made against you is to say that, if you were to put your degrees of belief into practice, you could be turned into a money pump (I'll explain what a money pump is forthwith). If I assign, say, a .6 probability to the proposition that it will rain today and also a .6 probability to the proposition that it won't rain today, I haven't straightforwardly contradicted myself. It's not clear what the world would have to do to make either of these statements false.

The problem emerges when I realize that, by my own lights, I should be willing to pay $6 for a bet that pays $10 if it rains (and nothing if it doesn't), and I should be willing to pay $6 for a bet that pays $10 if it doesn't rain (and nothing if it does). At the end of the day, if I've taken both of these bets, I'll have spent $12, and I've gotten back $10, no matter what happens out in the world. It does seem like a kind of failing of rationality if acting on my beliefs would guarantee that I would lose money no matter what happens to happen in the world. So, that's the Dutch Book. That's what it is to be "money pumped"; it's to be assured of losing money no matter how contingent events go.

This is supposed to be a theoretical problem, not a practical one. The idea is not that there's a Dutch bookie on every corner, waiting to take your money. You might be morally opposed to gambling, or you might just get suspicious (something might sound funny) if this sequence of bets is offered to you. So, it's not a worry about your financial future here. It's the fact that your degrees of belief are such that, if you were to act on them, you'd be assured of losing money—that's supposed to convince you that you're behaving irrationally.

It can be shown, mathematically, that if your degrees of belief obey the probability calculus (that's what follows from the axioms of probability we discussed last time), then no Dutch Book can be made against you. It can also be shown that any violation of the probability calculus makes you vulnerable to a Dutch Book. So, having your degrees of belief obey the probability calculus is a necessary and a sufficient condition for avoiding a Dutch Book.

So far, this is just a slightly more detailed version from a different angle of the stuff with which we ended last time. It is swell to have probabilistically coherent beliefs, but that requirement still permits intuitively crazy beliefs (like paranoid delusions), and none of this seems to have very much to do with science so far. Let's start fixing that. We want to make this a theory of scientific rationality, not mere coherence.

We're going to pay special attention to how we get a theory of scientific rationality by imposing, really, very modest additional constraints beyond mere probabilistic coherence.

Back in Lecture Twelve, which concerned such matters as the Raven Paradox, we saw Karl Hempel struggling to characterize a qualitative notion of confirmation. He wanted to talk about what it is for a piece of evidence to confirm a hypothesis. The question of how much confirmation the evidence provided for the hypothesis is supposed to get answered later. We're just tackling the basic question, Hempel thinks, of whether something is evidence for something else.

The Bayesians think this is a mistake. For a Bayesian, the notion of confirmation or evidence is quantitative right from the start. We can't ask whether a given bit of evidence confirms a hypothesis unless we know how probable the hypothesis started out being. So, in other words, in order to confirm a hypothesis, it has to start off with a prior probability.

A piece of evidence confirms a hypothesis just in case the evidence raises the *prior probability* of the hypothesis. This means that the probability of the hypothesis, given the evidence, is higher than the probability of the hypothesis had been without the evidence, or before the evidence came in.

Similarly, a piece of evidence disconfirms a hypothesis if the probability of the hypothesis, given the evidence, is less than the prior probability of the hypothesis had been.

All of this is done within this subjectivist or personal interpretation of probability; we're talking about what counts as evidence for you. So, a big dark cloud on an otherwise clear horizon counts as evidence of rain for me, just in case my subjective degree of belief that it will rain—given the new information that there's a big, dark cloud on the horizon—is higher than my prior probability that it would rain had been.

In saying this, I've tacitly invoked a notion called *conditional probability*, the probability of a hypothesis conditional on—or as I'll often say, given—the evidence. Here comes a mathematical definition of conditional probability. I'm stating it for the record because we're going to derive *Bayes's Theorem* from it (we won't do the derivation, but it's important for deriving Bayes's Theorem; it's not important that you memorize it, or sort of "get" the whole thing as it's being stated).

So, the conditional probability of a hypothesis, given the evidence, is the probability of the hypothesis and the evidence together, divided by the probability of the evidence (provided the evidence has a non-zero probability; we don't want to be dividing by zero). The probability of the hypothesis and the evidence together is the intersection of those two sets, as it were. It's the overlap of cloudy days and rainy days. The definition of conditional probability says that the higher the percentage of cloudy days that are rainy, the higher the conditional probability of the hypothesis that it will rain on the evidence that there's a dark cloud on the horizon. So, this is just a mathematization (I think that's a word) of a pretty intuitive notion. Let's connect this to the raised probability conception of confirmation.

Had I already been convinced that it would rain (because of the weather report, or because the voices in my head told me so), then this high conditional probability of rain, given the cloud on the horizon, would not raise my prior probability, and so wouldn't count as evidence for me of rain. It's on the assumption that I had been relatively neutral that the new information (that there's a big, dark cloud on the horizon) would confirm by raising the prior probability of the hypothesis that it would rain.

Let's briefly note that the idea that whatever raises the probability of a hypothesis confirms that hypothesis has some counterintuitive consequences. If I see Robert De Niro on the street, that probably raises, slightly, my subjective probability that he and I will make a movie together, but it hardly seems to count as evidence that we're going to make that movie. Maybe it's a teeny-tiny bit of evidence, but we're going to set such objections aside. It's quite common to think that for a bit of evidence to bear on a hypothesis positively is to confirm it. So, the raised probability conception of evidence is quite common, and I'm going to assume it, but I want you to note it's not completely uncontroversial or unproblematic.

So, that's the first thing we add to the notion of probabilistic coherence is a theory of evidence as probability raisers.

The second thing we need to add is that beliefs should be updated in accordance with something called Bayes's Theorem. Thomas Bayes was an English clergyman who proved this theorem in the 18th century. The theorem is a straightforward consequence of the definition of conditional probability, along with the definitions that make up the probability calculus that we talked about last time. It's an uncontroversial mathematical result. Non-Bayesians accept Bayes's Theorem just as surely as Bayesians do. You can apply it using a frequency theory of probability. The truth of Bayes's Theorem is not at issue; it's the use of Bayes's Theorem that makes a Bayesian a Bayesian.

Under the Bayesian interpretation, the theorem tells us how we are to update our beliefs in the light of experience, and it's this idea that's going to turn Bayesianism from a kind of unassuming theory of mere probabilistic coherence into a somewhat promising theory of scientific reasoning. These thin constraints on our subjective probability assignments (that they just have to cohere across time) get some teeth as we impose a requirement of probabilistic coherence not just at a time—but across time. I'll explain that momentarily.

First, let's just get the classic statement of the theorem. You don't have to take this in all at once; I'll be drawing on this repeatedly throughout the lecture.

So, the probably of a hypothesis, given the evidence (that's just the left-hand side, the conditional probability of the hypothesis on the evidence; we're defining that notion) is equal to the probability of the evidence, given the hypothesis (that's in the numerator of the right-hand side of the equation) times the prior probability of the hypothesis, and those two things are divided by the probability of the evidence. That's the theorem. If you've got your booklet, it'll be lovely to have in front of you. If you're driving your car, don't worry about what I just said.

The left-hand side of the equation, we've already talked about. That's just a conditional probability. But it can be given two different readings, depending on the time at which we're considering things. It depends on whether the evidence counts as already "in" or not.

If the evidence isn't yet "in," we interpret the left-hand side—the probability of the hypothesis, given the evidence—as the prior conditional probability of the hypothesis on, or given, the evidence (and the agent's background beliefs count in here too; I'm sort of oversimplifying by leaving those out).

So, it's what your current beliefs are about how probable the hypothesis would be if a certain piece of evidence were to come in. So, if I were a physicist in 1915 or so, I might have assigned a low probability (subjective probability) to Einstein's hypothesis of general relativity, but I also might have thought to myself, "If it turns out that light rays are bent by the sun, I will end up assigning quite a high probability to Einstein's hypothesis of general relativity." We don't generally go around declaring our prior conditional probabilities like that, but that's the kind of probability we're talking about, if the evidence is not yet in.

If the evidence is in, then the number on the left-hand side is called the *posterior probability of the hypothesis*—it's the probability that you assign to the hypothesis now that the evidence has come in. Once you discover that light rays are bent by the gravitational field of the sun, what probability do you, in fact, assign to the hypothesis? That's the simple part, the left-hand side.

We turn now to the right-hand side of the equation. The probability of the evidence, given the hypothesis, is a measure of how unsurprising the evidence is (if it's in) or would be (if it's not yet in), given the hypothesis. So, given Einstein's hypothesis of general relativity, the probability that light rays are bent by the sun's gravitational field is quite high. The probability of the evidence on the hypothesis is quite high.

That gets multiplied in the right-hand side of the equation by the prior probability of the hypothesis, however probable a given person thought Einstein's theory of general relativity was.

Since these numbers are in the numerator of the fraction, the posterior probability (it could also be the prior conditional probability; there are two interpretations of that—I'm just focusing on one for convenience's sake) goes up when these go up. The more expected the evidence would be, given the hypothesis, and the more likely you thought the hypothesis beforehand, the more likely you're going to find the hypothesis afterwards.

The prior probability of the evidence, apart from any hypothesis, is in the denominator of the fraction. This reflects the fact that, all other things equal, unexpected evidence raises posterior probabilities a lot more than expected evidence does. Apart from Einstein's theory, nobody would have thought the probability of light being bent by the gravitational field of the sun was very high. It's because Einstein's prediction is so unexpected, except in the light of Einstein's theory, that the evidence had so much power to confirm Einstein's theory.

That's the heart of Bayes's Theorem right there. The more unexpected a given bit of evidence is (by one's own subjective lights, of course) apart from a given hypothesis, and the more expected it is according to the hypothesis, the more confirmation the evidence confers on the hypothesis. So, your hypothesis is doing well if it makes something otherwise really quite surprising quite unsurprising. That's the heart of Bayes's Theorem, and this is just a mathematical representation of that fact that turns out to have some other uses as well. But it's not that hard.

The controversial part arises when the Bayesian proposes, as a rule of rationality, that once the evidence comes in, the agent's posterior probability for the hypothesis, now that the evidence is in, should be equal to the agent's prior conditional probability for the hypothesis, were the evidence to come in. That's a mouthful, but let me English it up.

It sounds like it should be uncontroversial. These are just the two different ways of interpreting the left-hand side of the equation. If the evidence isn't yet in, it's the prior conditional probability. If the evidence is in, it's the posterior probability. The Bayesian says your posterior probability, once the evidence comes in, should be the same as your prior conditional probability, were the evidence to come in.

What's controversial about that? Well, the math won't get you this result. This is an additional requirement. Because there are other ways I could maintain probabilistic coherence. Once I get the results of the Eddington eclipse experiment—that light is, in fact, bent by the rays of the sun—I could maintain my probabilistic coherence by altering some of my other subjective probabilities. I could decide I don't like that Einstein guy, let me tweak some different numbers on the right-hand side of the equation. I could decide, well, this evidence wasn't so surprising after all.

So, I can make my posterior probability—now that the evidence is in—different from my prior conditional probability, what I had said I would think were the evidence to come in. It's a substantive claim that those two things have to be identified. It's a substantive claim that today's priors should be the same as tomorrow's posteriors, if the evidence comes in. You don't get that for free.

The Bayesian has an argument, however. It's usually called the *diachronic* (meaning over time, rather than at a time) *Dutch Book Argument*. Our first Dutch Book Argument was synchronic; it was at a time. If your degrees of belief are incoherent at a time, you can be offered a series of bets that will guarantee that you'll lose money. A similar argument can be run across time. If you use a single rule for updating your beliefs (it's got to be a rule, not a whole bunch of incoherent principles), then a bookie who knows what rule you use can offer you a series of bets across time (some of these bets will depend on your future degrees of belief) and if you act on your present probability assessments and your present conditional probability assessments, then you'll be guaranteed to lose money. So, it's a Dutch Book Argument that considers you at various times. If you update your beliefs in any way other than by Bayesian conditionalization, you'll be guaranteed to lose money.

This is not generally thought to be as airtight a Dutch Book Argument as the synchronic one, but it's non-trivial. It does suggest there's a kind of rationality across time that Bayesian conditionalization gets you.

If we put all this together, we get an interesting approach to scientific reasoning. The Bayesian approach allows for an impressive range of subjectivity (there are very few constraints on prior probabilities; they have to cohere with your other degrees of belief), but also impressive objectivity (there is exactly one correct way of updating your beliefs in the light of new evidence).

So, this is a view that has the potential for incorporating a great deal of the literature on confirmation that we've seen earlier in this course. So, for instance, it can capture a central part of what the logical positivists had wanted, as well as a central part of what Kuhn—the great critic of the positivists—had wanted.

Even in cases of revolutions, it could be the case that all of these scientists are following the same scientific procedure (namely, Bayesian updating of prior probabilities), but the deep disagreements about whether we should be Newtonians, or Einsteinians, or whatever, are explained by plugging in different probability judgments into the very same equation. So, Bayesianism has room for the rules of the positivists (a precise, logical, mathematical rule about what rationality involves), but also for Kuhn's values, which are much more permissive, permit much more disagreement, while allowing everybody to meet standards of scientific rationality.

So, there's a sense in which Bayesianism offers a bit of commensurability across scientific revolutions, though perhaps not as much as the positivists would have wanted. It offers the prospect that scientific disagreement can be rational and discussable because we just plug the right values into Bayes's Theorem, while also making room for lots of idiosyncratic views about what the world is like and how a given bit of evidence bears on a given theory. You and I have the same conception of rationality, but we start from different places and have different conceptions of how the evidence bears on our theories.

Bayes's Theorem also seems kind of anti-metaphysical in a way that would have pleased both the positivists and Kuhn. The kind of rationality and scientific norm defended is coherence. There's no direct argument that a coherent set of beliefs will get the world right (this notion of "the way the world is" seemed metaphysical to Kuhn and to the positivists), but there does seem to be something scientifically valuable about updating your beliefs in a coherent way as new evidence comes in. And if you do that, arguably, you've lived up to all the scientific norms that really matter.

So, we've really got parts, at least, of the best of both worlds going here in this pretty simple mathematical setup. For reasons like these, Bayesianism helped rekindle interest in issues about evidence and justification. After the demise of the positivists and their theory of inductive logic, people kind of shied away from these issues.

We need to delve a bit more deeply—though you'll be pleased to know, no more mathematically—into Bayesianism in order to see how it helps us get going on these problems about evidence and justification.

While allowing that there's a sense in which any set of beliefs that maintains probabilistic coherence across time (not just at each time—you can do that without using Bayesian conditionalization; the idea is you have to maintain coherence across time; that's the diachronic Dutch Book)—the orthodox Bayesian says—is, in an important sense, beyond epistemic reproach. Such a person is doing everything that rationality requires.

Nevertheless, despite all that subjectivity, Bayesians think they can account for scientific agreement. They argue that the initial subjectivity starts to disappear when enough good evidence comes in. This is called the *washing out, or swamping, of prior probabilities*. It can be established that no matter how great the disagreement between two people is—there is some amount of evidence that will bring their posterior probabilities as close together as you would like. So, no matter how much you and I start off by disagreeing, if the right quality evidence comes in, it can be mathematically shown that we will come very, very close to agreeing. That's an impressive result, but it's subject to some significant limitations.

For one thing, if either of us assigns a prior probability of zero to a hypothesis, no evidence will ever increase that probability. It's in the numerator of the fraction; you multiply by zero, you're going to get zero. So, Bayesians add this pretty small, uncontroversial additional requirement that it's not okay to assign a probability of zero to any contingent proposition, any proposition that could possibly be true. It's a modest additional restriction on rationality that they impose, and we won't worry about that one.

We should also note that there is no assurance that this convergence will happen in a reasonable amount of time—in the long run, we're all dead. The amount and kind of evidence it would take to overcome some degrees of subjective disagreement is really quite epic. So, we don't want to oversell the washing out of the priors.

More importantly, the washing out results require that the agents agree about the probabilities of all the various pieces of evidence, given the various hypotheses in question (so they have to agree in all of their judgments about how unsurprising each hypothesis renders each bit of evidence). And that's a pretty impressive requirement of agreement to impose on two people who disagree fairly massively.

It suggests, unrealistically, that people can really disagree about the probability of hypotheses while agreeing about the probability of evidence bearing in particular ways on the hypotheses. But it might not be as bad as it sounds. In many cases, it's really quite clear how probable a given hypothesis renders a given bit of evidence. Typically, one might think a hypothesis (along with auxiliary hypotheses; we never escape this Quinean problem of holism we saw way back when) often entails the evidence. If my hypothesis is true, a given bit of evidence must be true. If my hypothesis is "All copper conducts electricity," and my evidence is "This is a bit of copper," the probability of that evidence on my hypothesis (of it conducting electricity) is 1. So, those are often really well behaved probabilities. You can get more agreement than you might think about how different hypotheses are borne on by evidence (sorry, it's a mouthful).

We can further examine Bayesianism's credentials as a theory of scientific inference by applying it to a couple of our old friends: the new Riddle of Induction and the Raven Paradox, a couple of our brain-breakers from the first third of the course.

Goodman's new Riddle seems like much less of a threat to the Bayesians since, at least for most of us, the prior probability that the next emerald observed will be "green" is much higher than it is for the hypothesis that the next emerald observed will be "grue." So, at any time before January 1, 3000, the other parts of the equation will function identically—each hypothesis confers a probability of 1 on finding a green or a grue emerald. Before January 1, 3000, the emeralds look the same. The denominator is the probability, independently of either hypothesis (of the hypothesis that all emeralds are green, and of the hypothesis that all emeralds are grue) of the next emerald being green or grue—which, again, is identical up until January 1, 3000.

So, every number in Bayes's Theorem is going to be the same except for the prior probability of each hypothesis. And since you and I attach a much higher prior probability to the hypothesis that all emeralds are green than we do to the hypothesis that all emeralds are grue, we will continue—as new evidence comes in—to attach a much higher probability to the hypothesis that all emeralds are green than that all emeralds are grue, and we do so quite rationally.

Since the probability of getting a green or a grue emerald is pretty high (once we've got an emerald, we both expect that it's going to look a certain way), there's not much confirming power. That's the number in the denominator. If the probability of getting the evidence, by my lights, is reasonably high, it doesn't confirm my hypothesis very much. But it doesn't confirm either hypothesis very much, and so it's only the prior probability of the hypotheses that are going to do any work here.

So, the Bayesian is going to say that there's nothing the matter with either inductive argument. It's fine to infer from the grueness of emeralds to their continued grueness, or from the greenness of emeralds to their continued greenness. But whichever hypothesis you think had been more probable going in will remain more probable going out, and so we can preserve our confidence, our rational confidence, in the hypothesis that all emeralds are green. If you're a grue speaker or a very peculiar green speaker, you might have gone in with a higher probability that all emeralds are grue than green—in which case, you'll come out with that probability. But it's not clear that that's a mistake.

Bayesians can handle the Raven Paradox more or less equally straightforwardly. As we noted back in the initial the discussion of the paradox, it's easier to swallow the idea that a white shirt confirms the statement "All ravens are black" if we make room for a quantitative—rather than a qualitative—notion of confirmation. Hempel didn't do that when he first formulated the paradox. That allows us to say that a white shirt does confirm "All ravens are black" (there was some convoluted logic behind that that I won't march us through again), but we might be able to live with that if the evidence of a white shirt confirms the hypothesis that "All ravens are black" much less than a black raven confirms "All ravens are black." It doesn't seem like such a difficult pill to swallow at that point.

If we look at Bayes's Theorem, we'll see that the greater the ratio of the probability of the evidence—given the hypothesis—to the probability of the evidence, the greater the power of the evidence to confirm the hypothesis. Once more with feeling, that just means the more expected our hypothesis makes the evidence, and the more unexpected the evidence, the better things look for our hypothesis.

That's the source of the difference in the confirming power of white shirts and black ravens to confirm "All ravens are black." The probability that the next raven I see will be black, given that all ravens are black, is 1. That's the probability of the evidence, given the hypothesis. That's much higher than the probability that the next shirt I see will be white, given that all ravens are black. That hypothesis does not make that evidence particularly likely. It's pretty much just my prior probability that the next shirt I see will be white.

That's the heart of the matter right there. That's the difference between the confirming power of a black raven with respect to the hypothesis that "All ravens are black" versus that of a white shirt with respect to "All ravens are black."

So, these are just some small illustrations of how Bayesianism has some real potential, starting from really quite modest resources, to become a plausible theory of scientific reasoning. It captures much of what Kuhn had wanted, much of what the positivists had wanted, and it looks like it can be applied to cases in a way that's simultaneously illuminating and somewhat plausible.

Next time, we'll start looking at some criticisms of the Bayesian approach.

Lecture Thirty-Two
Problems with Bayesianism

Scope:

Predictably, a Bayesian backlash has also been gaining momentum in recent years. This lecture investigates Bayesianism's surprisingly subjective approach to probability assignments, as well as the Bayesian treatment of the problem of old evidence (it appears that we can never learn anything from evidence that is already in). We compare the Bayesian approach with competing conceptions of statistical inference, such as those derived from classical statistics. This assessment results in a cost-benefit analysis rather than a vindication or a refutation.

Outline

I. Though mathematically intensive, the basic ideas behind Bayesianism are rather simple and powerful. It has gained many adherents in recent decades, but with increased attention has come increased criticism. We begin with a couple of criticisms that we will not pursue in detail.

   A. Bayesianism involves a rather dramatic idealization of human cognizers.
   1. We do not have the processing power to meet Bayesian standards even in fairly simple cases. Coherence requires logical omniscience, namely, that we know all the logical consequences of our beliefs, and that is unrealistic.
   2. On the other hand, it is not clear how descriptively accurate the theory needs to be. There is at least some role for ideals that cannot be met.

   B. Many think that Bayesianism does not reflect actual scientific practice. Scientists do not think of their work in terms of degrees of belief. They leave themselves out of the picture when doing science.

II. The *problem of old evidence* represents a longstanding challenge to the Bayesian approach.

A. It seems that scientific theories can be confirmed by facts that are already known. Newton's theory, for instance, could explain Kepler's well-known laws, and thus, Kepler's laws are evidence for Newton's theory.

B. But any evidence that is already known for sure should, it seems, receive a probability of 1. And because $P(E)$ is 1, $P(E/H)$ is 1.

C. If we plug these numbers into Bayes's Theorem, we quickly see that old evidence has no power to confirm hypotheses. The prior probability of the hypothesis is multiplied by 1 over 1, so the posterior probability stays equal to the prior probability.

D. A couple of responses are open to the Bayesian.
   1. Bayesians can claim that the subjective probability of E should not be considered against one's actual background knowledge (because that knowledge includes E) but, instead, against what the background knowledge would be if E were not yet known. This involves ascertaining how surprising E would be if neither it nor anything that entails it were included in our background knowledge.
   2. But the counterfactual "how surprising would I judge E to be if I did not already know it" can be difficult to evaluate once we realize how many statements might bear on E.
   3. Alternatively, the Bayesian can say it's not really E that confirms H when E is already known. In the Newtonian case, it was the new information that Newton's theory entailed Kepler's laws that did the confirming. It is "H entails E," not E, that confirms.
   4. This involves a couple of problems: Sometimes, the fact that H entails E itself seems like old evidence. And anyway, mightn't we want to insist that E confirms H in such circumstances?

III. The most influential objections to the Bayesian program concern its somewhat brazen tolerance of subjective probabilities. In scientific contexts, Bayesianism can be supplemented with various strategies or scientific values designed to impose constraints on admissible probability of conditional probability

assignments so that outlandish degrees of belief are regarded as legitimately criticizable. The extent to which subjectivity can be tempered is different for the various terms of Bayes's Theorem.
- **A.** P(E/H) is usually pretty well behaved. Often, the hypothesis in question entails the evidence, in which case P(E/H) is 1. The probability of this piece of copper conducting electricity given that all copper conducts electricity is 1.
- **B.** The prior probability of the hypothesis can be tamed a bit.
    1. One might try to impose norms requiring one to look for evidence (for example, observed frequencies) relevant to setting prior probabilities, and one might evaluate new hypotheses by comparing them to similar hypotheses.
    2. This is trickier than it sounds. What is to count as a hypothesis that is similar to the one in question? Evidence seems unlikely to settle the relevant similarity relation.
- **C.** The hardest problem concerns the denominator of Bayes's Theorem—P(E).
    1. The probability of the evidence is equal to the probability of the evidence on the assumption that our hypothesis is true plus the probability of the evidence on the assumption that our hypothesis is false: P(E) = [P(E/H) × P(H)] + [P(E/~H) × P(~H)].
    2. The claim that our hypothesis is false is not itself a hypothesis. It is called the *catch-all hypothesis*. There are endless ways in which our hypothesis could be false, and they will not all assign the same probability to the evidence.
    3. The only way to get solid evidence for the values in this part of the equation is to claim that all the possible hypotheses are under consideration. And this is generally not warranted.
    4. Thus, even if you can temper or constrain many of the probabilities that figure in the theorem in the light of evidence and scientific practice, there is no getting around the fact that one of the probabilities in the equation can be only a kind of guess about how surprising a certain result would be. Subjectivity of this

sort cannot be eliminated if you're going to use Bayes's Theorem.
- **D.** Though many scientists think that nothing having anything to do with subjectivity should be let anywhere near science, it's not clear how bad the subjectivity built into the Bayesian program is. Kuhn, for instance, thought a limited role for subjectivity was crucial to the health of science.

IV. Classical statistics tries to avoid this role for subjectivity. It remains dominant in most scientific disciplines, though Bayesianism is on the rise.
- **A.** If you don't start from probabilities for hypotheses, you can't end up with them. Because non-Bayesians don't like the role of subjectivity in setting $P(H)$, they forego getting any values for $P(H/E)$.
- **B.** Basically, that leaves them with $P(E/H)$ doing most of the work.
  1. In classical statistics, we want to know whether a given correlation is significant or random.
  2. We start by assuming that the results are random, and we run some tests.
  3. Very roughly, if $P(E/H)$, that is, the probability of getting results like these randomly is lower than .05, then we reject the assumption of randomness and call the correlation significant.
  4. This significance threshold is both somewhat arbitrary and somewhat sacred in science. One might ask exactly why rules that are arbitrary but objective are better than probability judgments that are subjective but objectively updated; however, we can only touch on these issues here.
  5. The 5% threshold has occasionally been abused or applied mindlessly, and that has led to some very questionable scientific work.
- **C.** Another statistical confusion looms large in public discussions of evidence. All sides agree that it is important to keep $P(E/H)$ quite distinct from $P(H/E)$, but people make this mistake quite often.
  1. In a criminal trial, the jury might be told an impressive $P(E/H)$, for instance, that the forensic evidence matches

the defendant and that the chances of its matching a person chosen at random are miniscule.
2. This evidence has a lot of potential power to confirm the hypothesis that the defendant is guilty, but it cannot do so without a prior probability for that hypothesis. If I were having dinner with the defendant 1,000 miles away from the scene of the crime, the evidence will rightly fail to sway me. It's only by assuming certain values for the prior probability of H that this argument goes through.
3. Clearheaded approaches in the tradition of classical statistics understand that one can never get directly from P(E/H) to any value for P(H). Recent developments in the field try to allow the classical approach to do a lot more than apply mechanical rules of rejection, while still avoiding a role for prior subjective probabilities. But they cannot provide posterior probabilities, which may be a bug or a feature.

Essential Reading:

Glymour, "Why I Am Not a Bayesian," in Curd and Cover, *Philosophy of Science: The Central Issues*, pp. 584–606.

Supplementary Reading:

Kelly and Glymour, "Why Probability Does Not Capture the Logic of Scientific Justification," in Hitchcock, *Contemporary Debates in Philosophy of Science*, pp. 94–114.

Questions to Consider:

1. You and I are not capable of meeting the demands of Bayesian rationality—we simply don't possess adequate computing power. Is this an objection to Bayesianism as a theory of scientific reasoning or not? Is it an objection to a moral theory if you and I aren't capable of living up to its demands? Why or why not?
2. What would be the pros and cons of instructing jurors to think of themselves as updating prior subjective probabilities on the basis of the evidence presented to them?

Lecture Thirty-Two—Transcript
Problems with Bayesianism

Though the math can be a bit daunting, the basic ideas behind Bayesianism are rather simple and powerful. We've seen that it strikes an interesting balance between a role for subjectivity and a role for objectivity in scientific reasoning, and it has at least some potential resources for handling a lot of problems about confirmation that we weren't able to handle until we brought in notions like probability. Bayesianism has gained many adherents in recent decades, but with the increased attention has come increased criticism, to which we now turn our attention.

We begin by noting a couple of objections to the Bayesian program that are interesting, but that we won't be able to pursue in detail.

One is that it involves a rather dramatic idealization of human cognizers. The way I've presented it so far, I've made it sound kind of modest to talk about merely requiring coherence of one's degrees of belief, rather than that one get the world's relative frequencies right, or meet some a priori logical principle of indifference. Nevertheless, mere coherence—especially when probability gets brought to the table—is, in many ways, enormously demanding.

You and I do not have the processing power to meet Bayesian standards of probabilistic coherence, even in fairly simple cases, much less in cases of serious scientific complexity. Moderately precise degrees of belief plus the demand for coherence would break our brains in nothing flat. Coherence requires what philosophers sometimes call *logical omniscience*, that we know all the logical consequences of our beliefs, and if you impose probability on top of that as a requirement to see how our beliefs hang together, the amount of processing power required is just completely beyond us.

On the other hand, it's not clear how descriptively accurate the Bayesian conception of scientific reasoning needs to be. Surely, there is some role for ideals that cannot be met by you and me. I probably can't meet the standards of deductive coherence either. There are probably always going to be some contradictions buried somewhere in my belief set. That doesn't undermine the status of the norm that you should try not to believe contradictions. So, even though you and I can't begin to meet Bayesian standards of scientific rationality, that doesn't automatically disqualify them as appropriate standards of scientific rationality.

This raises the question of what kind of theory Bayesianism is best posed as. To what extent should it model itself on the kind of rational reconstruction approach we saw from the positivists, and to what extent should it market itself as a kind of Kuhnian descriptive theory about how well-conducted scientific reasoning actually goes.

So, along these lines, there is an objection with a Kuhnian flavor that has a bit of force. Many people think that Bayesianism gets in trouble because it does not reflect actual scientific practice, and in noteworthy and important ways. You don't generally find scientists reporting their results by stating their prior probability assignments and then showing how the evidence convinced them to raise or lower their probability assignments. Most scientists don't think of their work in terms of degrees of belief at all. They're not sharing some autobiographical information about how their beliefs have been updated. This is a broadly Popperian observation: Belief doesn't seem to have the kind of role in science that the degrees-of-belief approach of the Bayesians suggests it should have. We'll come back around to this at the end of this lecture.

But for now, we can note that it remains the case that relatively few scientists think of themselves as Bayesians, but the ranks of Bayesian scientists seem to be growing, and that has some force against this Kuhnian objection that it doesn't get scientific practice right. And, of course, to some extent, one need not explicitly use a theory (or think of oneself as using a theory) to count as deploying the theory. It's the apparent tension between scientific practice and what Bayesianism says scientists are doing, not the mere fact that scientists don't explicitly use Bayesianism—that's the apparent problem, but there's some resources Bayesians can use to try to narrow the gap there.

We turn our attention now to more directly epistemic problems with Bayesianism, and these we will talk about in a bit of detail. The *problem of old evidence* is perhaps the most longstanding worry lodged at Bayesians. As we noted a couple of lectures ago, it seems that scientific theories can be confirmed by facts that are already known, by data that's already in. Newton's theory, for instance, could explain Kepler's well-known laws of planetary motion; it could explain well-known facts about how the tides behaved, et cetera, and that all counted as evidence for the truth of Newton's theory.

But one of Bayesianism's strengths starts to look like a weakness here. Because, as we've seen, the fact that the probability of the evidence, apart from any hypothesis, is the denominator of Bayes's Theorem means that unlikely evidence, surprising evidence, has much more confirming power than does expected evidence. This is an insight associated with Popper, who thinks that scientists should be putting forward bold, surprising conjectures, and it's borne out in things like the eclipse experiment that confirmed Einstein's theory.

But this feature of Bayesianism, that it captures nicely this idea, can also be a bug because it looks like it deprives any evidence that's already known (any evidence, at least, that's already known for sure) of having any power to confirm at all. Why? Because if you know a bit of evidence for sure, then the probability of that evidence, by your lights, is 1, as is the probability of getting that evidence, given any particular hypothesis. If I assign a probability of 1 to "I'm staring at a camera right now," then given any hypothesis, I'd better assign a probability of 1 to "I'm staring at a camera right now." So, the probability for me that I'm staring at a camera right now—given the hypothesis that donkeys fly—is 1. So, both the probability of the evidence and the probability of the evidence on any given hypothesis is 1.

If you plug those numbers into Bayes's Theorem, you'll see that old evidence has zero power to confirm hypotheses because the prior probability is getting multiplied by 1 over 1. So, if the evidence has a probability of 1, your posterior probability is guaranteed to stay the same as your prior probability—old evidence cannot confirm a hypothesis in the slightest.

That looks like a problem because we thought that Newton's theories were confirmed by data that it explained and accounted for, that had already been known. So, there are a couple of strategies open to Bayesians.

One is that they can claim that the subjective probability of the evidence is not to be considered against one's actual background knowledge of the world. Why? Because that background knowledge includes the evidence. You already know the evidence. So, instead, what you're supposed to do is assess the probability of the evidence using your background knowledge if the evidence were not already known. We're supposed to ascertain how surprising the evidence would be if we didn't already know it, or anything that entails it. So, it's a counterfactual conception of how surprising the evidence would be.

As you might imagine, this is not an easy counterfactual to evaluate because the evidence could be integrated into our web of belief in all kinds of ways. So, how exactly am I supposed to arrive at a clear, confident answer to the question: "How surprising would I judge this evidence to be if I didn't already know it?" because the number of statements that could bear on E (the evidence statement) in various ways is really quite enormous. So, that's a hard counterfactual to evaluate. It's not clear that I can give a clear answer to the question: "How surprising would I think this evidence about the tides were if I didn't already know it?"

So, another route open to the Bayesian is to say that it's not the evidence that really confirms the hypothesis when the evidence is already in; it's the fact that the hypothesis implies the evidence. So, in the Newtonian case, it's the new information that Newton's theory entails Kepler's laws that confirms Newton's theory. So, it's "H entails E"; the hypothesis entails the evidence, rather than the evidence statement that's doing the work.

This might work, but it has a couple of problems. For one thing, sometimes the fact the hypothesis entails the evidence is itself already known. So, the very same problem of old evidence applies to the fact that the hypothesis entails the evidence. And anyway, it's not clear that this answers our question. We thought that evidence that was already in could confirm a hypothesis, and being told that maybe the fact that the hypothesis entails the evidence can confirm the hypothesis, didn't answer our question, which is why, if at all, it is impossible for evidence that's already known to confirm a hypothesis. So, some think the Bayesian evades the question rather than answering it, if that reply is offered.

The most influential objections to the Bayesian program concern its somewhat brazen tolerance of subjective probabilities. We've emphasized this in our earlier presentations: The only rules are rules of coherence, at least initially, and then you have to update in the right sort of way. We've seen that the Bayesian can appeal to the washing out of the priors to argue that the subjectivity is not too damaging to serious science. If the evidence is good enough, then people will converge, and so we can see some of the kind of convergence that traditional views of science have wanted to credit scientific inquiry with attaining.

But those washing out results were themselves subject to pretty significant limitations about agreement on how evidence bears on theory. So, we don't want to just say "Subjectivity is fine because if perfect evidence comes in, and if two people agree on how the evidence bears on theory, they'll come to agreement." That still seems, to many people, to tolerate too much rampant subjectivity, to allow scientists to have initial probability assignments based on whim, or prejudice, or something like that—we want to try to rule those out, at least to some extent.

So, if Bayesianism is to be used as a theory of scientific reasoning (that's not the only use for it; it's a theory of everyday reasoning too), many people think it needs to make room for some distinctively scientific constraints on the values that get plugged into the equation. So, these distinctive scientific constraints are not part of orthodox Bayesianism, but they're not incompatible with it either. These are things we just add to try to make it a better theory of scientific reasoning.

The idea is to use evidence and scientific values to impose some substantive constraints on admissible probability or conditional probability assignments so that outlandish degrees of belief are regarded as legitimately criticizable. They don't violate Bayesianism, as such, but they violate a scientifically informed version of Bayesianism. This is often called *tempered personalism*; we require that scientists temper their subjective probabilities in the light of scientific norms and values. So, we can exclude some degrees of belief from consideration as appropriately scientific.

The extent to which parts of the equation can be tempered varies significantly. As we saw last time, the probability of the evidence, given the hypothesis, is usually pretty well behaved. Often, the hypothesis in question entails the evidence—in which case, the probability of the evidence—given the hypothesis—is 1. The probability of this piece of copper conducting electricity—given the hypothesis that "All copper conducts electricity"—is 1. Everybody can agree to that; that gets tempered really quite nicely. In other cases, the hypothesis might not entail the evidence, but in a well-set-up experiment, it confers a fairly definite probability on the evidence. That's part of why experimental setups are run the way they are: You can get a clear, logical relationship between the hypothesis and the evidence and know how surprising a given bit of evidence would be if the hypothesis were true.

So, the probability of the evidence, given the hypothesis, is not particularly problematic. Most people think that can be rendered reasonably tempered, reasonably objective.

We turn now to the other part in the numerator of the right-hand side of Bayes's equation, which is the prior probability of the hypothesis—and that can be tamed a good bit. Within science, anyway, we could try to impose norms requiring that one look for evidence, not just start from whatever whim one finds oneself with, evidence that seems relevant in ways guided by scientific practice to the setting of prior probabilities.

We could try to require that hypotheses put forward by serious scientists should not be given probabilities that are ridiculously low. We could also require that, since many hypotheses put forward by serious scientists turn out to be false, you shouldn't start with too high an initial probability assignment either—and try to independently defend some rules constraining the range of prior probability assignments.

So, this is part of a broadly naturalistic approach. We look at what science does and try to build those norms into our theory of scientific reasoning. All this is supposed to be an improvement on letting probability assignments get settled in any way that happens to cohere with other things the scientist thinks because a given scientist might just be in a bad mood and have very peculiar probability assignments for that reason.

Nevertheless, there is a limit to the extent to which the prior probability for the hypothesis can get tempered. Because I'm supposed to assign a prior probability to the hypothesis in the way that science would do so—I'm supposed to consider how similar hypotheses get treated in science. That's still supposed to allow some latitude for individual scientific disagreement, but it's supposed to be a significant constraint. But now we face the problem of what counts as "hypotheses like this one" if I'm to treat this as like hypotheses have been treated. Am I to treat all inverse-square law hypotheses similarly no matter what phenomena they're applied to? Am I supposed to treat all hypotheses proposed by scientists who got their Ph.D.s from MIT similarly? What counts as a similar hypothesis by the standards of science is going to need to be addressed, and it looks like there's going to be a lot of room for a kind of subjective probability in that assessment.

These aren't insurmountable problems, to the extent that you think you can articulate science's own standards of similarity, but that is a non-trivial problem.

So, the probability of the evidence, given the hypothesis, is pretty well behaved. The prior probability of the hypothesis is moderately well behaved.

The hard problem, surprisingly, concerns the denominator on the right-hand side of Bayes's Theorem, the probability of the evidence apart from any hypothesis. Why? Because if we unpack this explicitly, the probability of the evidence is equal to the probability of the evidence on the assumption that our hypothesis is true, plus the probability of the evidence on the assumption that our hypothesis is false. If we've got more than one serious hypothesis, then we divide it up: the probability of the evidence on the assumption that hypothesis 1 is true, plus the probability of the evidence that hypothesis 2 is true, plus the probability of the evidence on the hypothesis that neither is true.

So, let me unpack this with a little explicitness. You don't need to understand exactly how all of this works. I'll focus on the components we're going to need.

The probability of the evidence (I'm just going to use the one-hypothesis case, for simplicity's sake) is the probability of the evidence given the hypothesis times the probability of the hypothesis. So, that's how probable the hypothesis makes the evidence times how probable the hypothesis itself is.

Same thing if the hypothesis is false. The probability of the evidence if the hypothesis is false times the probability that the hypothesis is false. That tells me how probable my evidence is, period, since there are only two possibilities (that the hypothesis is true, and that it's false).

The components there—the probability of the evidence given the hypothesis and the prior probability of the hypothesis—aren't too troublesome; we just saw that. The problem is that the claim that our hypothesis is false is not itself a hypothesis. There are many, many ways for a hypothesis to be false. So, the probability that a given hypothesis is false is not a real hypothesis; it's called the *catchall hypothesis*. Because there are deeply different ways in which a hypothesis could be false, and they won't all assign the same probability to the evidence. The hypothesis could be false in such a way that the real world doesn't exist, we're all in the matrix, or it could be false because a competing hypothesis that's well-known to science turns out to be true. There are a gazillion ways for a hypothesis to be false. So, we're not talking about a particular hypothesis; we're talking about a catchall.

The only way to get solid evidence for the values in this part of the equation, in the denominator, would be to claim that all the possible hypotheses are under consideration, so that we can ignore the catchall hypothesis. All we need to consider is the probability of the evidence given that, say, the wave theory of light is true, and the probability of the evidence given that the particle theory of light is true. If those really exhausted logical space, we wouldn't have to worry about the catchall hypothesis.

Scientists often don't worry about the catchall hypothesis, which reflects their confidence that they have exhausted all of the serious possibilities. They think philosophers are only interested in weird possibilities like we're all dreaming, or we're in the matrix, or something like that. While philosophers generally want to focus on cases in which scientists in the 19th century all thought either the wave theory of light or the particle theory of light had to be true. It turns out they're both false—one of the gazillion hypotheses under the catchall turns out to be true.

So, it's logically illegitimate to ignore the catchall hypothesis, but there's no way to get evidence directly to bear on the catchall hypothesis because it's not a particular hypothesis. So, that part of Bayes's Theorem, the denominator on the right-hand side is one that can only be given a kind of subjective interpretation. So, this is one of those places where philosophers are not all that deferential to what scientists actually do because we think they generally make a logical mistake by ignoring the catchall hypothesis.

So, even if I'm able to temper or constrain many of the probabilities that figure in Bayes's Theorem in the light of evidence, in the light of scientific practice, there's no getting around the fact that the only way to use the whole equation is to include a kind of psychological guess about how surprising it would be to get this evidence if all of my seriously considered hypotheses are wrong. Subjectivity of this sort cannot be eliminated if you're going to use Bayes's Theorem.

This is often presented as a devastating objection to Bayesianism, but it's not clear how bad this subjectivity is. We've seen Kuhn arguing a few times that a certain kind of subjectivity is healthy for science; disagreement across scientific revolutions allows serious possibilities to flourish. We've seen other people argue that diversity of subjective views, under conditions of free discussion and other institutional supports about scientific criticism, can help increase the objectivity of science. So, the issue should be what's good or bad for science about the extent to which Bayesianism allows subjective probabilities.

In addition, Bayesians can argue that they're just making the role of subjective factors in science explicit—and hence, discussable. Other approaches, they say (and this will emerge more clearly in a few minutes), involve subjective factors, but tend not to admit it (subjectivity is really the "S-word" in science). So, let's turn to some major competitors with the Bayesian program and see if we can assess the costs and benefits of each approach.

Lots of statisticians, scientists, philosophers of science are anxious to avoid the mention of subjectivity—at all costs. So, they don't want to start from probabilities for any hypotheses that are not thoroughly well behaved. Non-Bayesians don't like the subjectivity in the prior probability of the hypothesis—and, as we just saw, in the probability of the evidence. So, they forego getting any posterior probability value. You can't get a posterior probability if you're not willing to plug in values for these parts of Bayes's Theorem that can only be ascertained via a kind of subjective guess.

So basically, that leaves non-Bayesians using the probability of the evidence—given the hypothesis—to do most of their work. That's the well-behaved probability; that's the thoroughly well-behaved one, at least in most cases. This kind of approach is often paired with a kind of frequency interpretation of probability because these people don't want to rely on potentially subjective degrees of belief in the first place. They want probability statements to be about something observable.

One way of doing this is to appeal to classical statistics. Versions of this approach dominated the sciences for most of the 20th century, and they probably still do—though I'm not sure about that; Bayesians have been making real inroads here. Speaking loosely, and in a way that I'll explain shortly, classical statistics is Popperian while Bayesianism is more Kuhnian. I'm going to breeze through the very basics of classical statistics. I'm not going to be able to do full justice to the view, but I think you'll get as much as is needed for our purposes.

As so often is the case in discussions of probability, we'll appeal to urns filled with marbles (very well-behaved cases). Suppose we know that one urn contains 75 percent blue marbles and 25 percent red marbles. The other urn has the proportions reversed.

We can afford to draw two marbles from one of the urns (this part of the model is supposed to reflect the fact that it costs resources to test hypotheses in real science). Let's say we draw two marbles, both of which are blue, and we want to try to tell from which urn the marbles were drawn. This is a very simple problem in classical statistics.

Classical statistics has you state a hypothesis. Presumably, you would guess that these marbles came from the urn that has 75 percent blue marbles. What are you supposed to do? You assume that your hypothesis is false. To do that is to adopt what they call the *null hypothesis* (that's a term used a few different ways, but that'll be okay for now).

Now we calculate how likely the evidence would be on the null hypothesis. How surprising would it be to get this evidence if our original hypothesis turns out to be false?

So, what's the probability of drawing two blue marbles from the urn that's mostly got red marbles in it? It's 75/25, so we have a 25 percent chance for each marble of having drawn it, if it's from the urn that's got 3/4 red marbles. So, it's 25 percent chance times a 25 percent chance, which comes to a 6.25 percent chance that these marbles came from the urn of mostly red marbles.

According to classical statistics, you would be allowed to reject the null hypothesis if the probability of that evidence on the null hypothesis is below 5 percent. It isn't in this case (it's 6.25 percent)—so, the result is not statistically significant, and so, you've learned nothing.

This is a little bit counterintuitive. A Bayesian, by contrast, would have appealed to subjective (at least in some sense) prior probabilities and would have updated beliefs in light of this modest—but intuitively genuine—information, and so would have gone in with some subjective degree of belief and modified it by taking this new evidence into account.

This is what I mean when I suggest that classical statistics is Popperian. We don't adopt hypotheses; we have rules for rejecting them. The world can tell us we're wrong in a way that it can't tell us we're right. That's a broadly Popperian approach.

This 5 percent significance threshold is very interesting. It is both somewhat arbitrary and somewhat sacred in science. There was not much argument offered for it. It was proposed in the 1920s by the British statistician R.A. Fisher. He pretty much just said that if you had a less than 5 percent chance of randomness producing results at least as impressive as the results you've observed—then your finding seemed to him to be significant. It's a mathematically convenient standard, and he thought it provided reasonable protection against error. That's not a trivial argument, but it's not exactly a knockdown argument. But the 5 percent significance threshold has been institutionalized, in scientific journals, in government agencies—it's the scientific gold standard, to a certain extent.

So, though it's semi-arbitrary, it has one thing going for it: It's—in an important respect—objective. The data plus this rule provides a decision procedure that leaves no serious room for judgment or disagreement. This is crucial to why it has caught on to the extent that it does, because it's not subjective; the S-word is banished.

It's replaced, one might think, by a somewhat arbitrary—though objective—rule, and one might wonder why arbitrariness is so much better than subjectivity. It's not that I think it's such an objectionable standard, but we should understand what it is. It is not God's, or science's, or the world's own standard for when you've got a good hypothesis. It's surprising how many people won't admit that one could do science differently because that seems to introduce the issue of subjectivity. We now have to admit that we have scientific values and decide whether the 5 percent threshold is the right one to apply.

In addition (this is somewhat controversial, but not very, I don't think), the mindless application of the 5 percent standard has sometimes led to some reasonably bad science. It doesn't often happen too vulgarly, but it's clear that the standard can be abused if it's combined with the wrong procedures. We learned back in our first discussion of the Raven Paradox that procedures can matter. The procedure of looking at something black and telling whether it's a raven is different from the procedure of looking at something that's a raven and telling whether it's black.

How does that apply here? If you run enough, more or less, thoughtless experiments—eventually, you're going to get some statistically significant findings. Whether there's any correlation or cause out in the world, if you perform the procedure of just running enough experiments and publish when you get a 5 percent threshold result, you're going to get some really bad results published out there. At least some people think that some of the real disappointments in biomedical research are when drugs that were "clinically shown" to be effective turn out to have little or no value. The culprit here is, arguably, a desire for objectivity run amok, a refusal to rely on judgment about whether the procedures are sensible. You just run procedures over and over (and again, I'm not accusing anybody in particular of doing this; science is generally better than this), but there's room for the standard to be abused.

Another statistical confusion looms large in public discussions of evidence outside of science as well as within it. All sides agree that there's a big difference between the probability of the evidence, given the hypothesis, and the probability of the hypothesis, given the evidence. But they are confused quite often.

Lawyers have a name for this; it's sometimes called the *prosecutor's fallacy*—because in a criminal trial, the jury might be told that there's a very impressive probability of the evidence, given the hypothesis. So, for instance, the forensic evidence matches the defendant's DNA, and the chances of the evidence matching a person chosen at random might be miniscule. So, on the hypothesis that the defendant did it, this evidence is unsurprising; and on the hypothesis that the defendant didn't do it, it's much more surprising.

That suggests that the evidence has a lot of potential power to confirm the hypothesis that the defendant is guilty.

You can see that if you plug those values into Bayes's Theorem, but it cannot confirm the hypothesis that the defendant is guilty, unless you've got a prior probability for that hypothesis. You cannot get to the probability of the hypothesis, given the evidence, without putting all those values, all the subjective values, into Bayes's Theorem.

We can see this by realizing that if I knew for sure I was having dinner with the defendant 1,000 miles away from the scene of the crime, the evidence—though it has, in an abstract sense, a lot of confirming power—will rightly fail to sway me because my prior probability for the hypothesis that the defendant did it will be low because I was having dinner with the defendant 1,000 miles away.

But lawyers don't want to come out and say that they're appealing to subjective degrees of belief on the part of jurors (and, of course, many lawyers aren't statisticians anyway). So, they can't quite come out and say, "Well, if you have ordinary probabilistic assignments here, then this evidence should lead you to assign a high probability to the hypothesis that the defendant did it." So, they talk as if the probability of the evidence—given the hypothesis—by itself can make the probability of the hypothesis—given the evidence—high. That's a logical fallacy. It will work if we assume some background beliefs, which people assume, but won't admit that they assume. That's a real problem in clear thinking about evidential matters.

Clear thinking would require that one live by the Popperian scruples of the classical approach to statistics, which admits that you can't get a probability for hypotheses. You only get a rule for rejecting. So, the fear of subjectivity muddles our thinking about a lot of matters here.

I should note, to give the classical tradition more credit than I have, it hasn't just been sitting on its hands while Bayesianism has been catching on. Classical views can adopt a lot of intellectual virtues; they can do more than just apply a mechanical rule of rejection. What they've done in recent decades, while avoiding prior subjective probabilities, they've tried to come up with a way of helping us learn more from experience. So, they cannot attach posterior probabilities to hypotheses; you have to have all of the values for Bayes's Theorem to do that.

But they can provide methods that have certain desirable characteristics. So, one classical approach within classical statistics allows you to distinguish between errors of two kinds. One would be accepting the hypothesis if it's false; the other would be rejecting it if it's true. You specify the maximum value you're willing to tolerate for one kind of error, and then the method will minimize the chance of error of the other kind. Or there's a kind of 20-questions method that will—if there's a yes-or-no answer to a question—give you one in the smallest number of steps possible.

The point is that the classical view has to admit that they cannot attach probabilities to hypotheses, but they want to go beyond Popper and do a better job of letting us learn from experience.

So, what we've got is a choice between two different sorts of views. If you're willing to fess up to a role for subjective prior probabilities, then it's straightforward that you can update your beliefs and count things as evidence for the truth of hypotheses. If you're not, you have to take a more direct approach. The fear of subjectivity costs something. And I'm not saying it's a mistake to try to banish subjectivity, but it's a mistake not to admit that it has costs. That's, I think, what vindicates this detour through so much mathematics.

We turn next time to philosophy of science within particular sciences, and physics is the first victim.

Lecture Thirty-Three
Entropy and Explanation

Scope:

Most philosophy of science these days is philosophy *of* a particular science and, more particularly, of a particular issue or theory within one of the sciences. As we wind down the course, I will try to offer some illustrations of how the general issues in philosophy of science that we have discussed are being treated within contemporary, relatively specialized philosophy of science. In this lecture, we turn to the philosophy of physics and examine an intriguing package that includes the reduction of thermodynamics to statistical mechanics, the direction of time, the origin of the universe, and the nature of explanation.

Outline

I. We turn now to a series of relatively detailed examinations of philosophical issues that arise within particular sciences. These both illuminate and are illuminated by the general philosophical issues on which we've so far focused.

   A. Our topic from the philosophy of physics can seem frivolous (especially the way I've chosen to express it), but it raises deep issues about explanation and reduction. Why can I not stir milk out of my coffee?

   B. In one sense, I *can* stir milk out of my coffee. The basic laws of nature (both classical and quantum mechanical) permit it. They permit all of the gas in a container to cluster in one corner, and they permit heat to flow from a metal bar that has been kept in the freezer to one that has been kept in a hot oven.

   C. There is nothing in the basic laws of motion specifying in which direction molecules must move—a reverse motion is permitted by the basic mechanics.

   D. But, though they are permitted by the basic laws—in this case, statistical mechanics—the laws that in some sense reduce to these more basic laws tell us that air never leaks into a punctured tire.

E. The second law of thermodynamics says that energy tends to spread out. Another way of saying this is that *entropy* (a measure of this dissipative tendency) tends to increase. A system that is energetically isolated (energy is neither added to nor removed from the system) will tend to move toward an equilibrium state—heat will spread out and stay spread out.

II. Why does the second law of thermodynamics hold, given that there is nothing about the laws of motion, taken just by themselves, to make it so? The 19th-century Austrian physicist and mathematician Ludwig Boltzmann worked out the two most influential answers to this question.

A. Boltzmann's first answer is that the effect of collisions between rapidly moving gas molecules would tend to bring about an increase in entropy until it reached its maximum value. As Boltzmann realized, his answer was statistical, not deterministic.

B. Boltzmann's second answer suggests that there are just more ways for particles to spread out than there are for them to be concentrated. This can be thought of as a matter of multiple realizability: high-entropy states are realized by many more lower-level states than low-entropy states are. This explanation also makes the second law statistical.

C. The first explanation provides a mechanism for the tendency for entropy to increase: the collisions bring about the increases in entropy.

D. The second approach explains without providing a mechanism. Just as you don't need a causal story about the shuffling of cards to understand why you never get dealt a royal flush, the causal details are largely irrelevant to the second explanation.

III. Whichever explanation we adopt, we will run into puzzles about the direction of time, however.

A. Let's take the second explanation first. Just as almost all of the states a closed system can move to are high-entropy states, just about all the states it could have moved from are high-entropy states. Thus, it would seem that entropy should increase as we move toward the past, just as it does when we

move toward the future. But that never happens. There is a *temporal asymmetry* at the observable, thermodynamic level that this explanation does not seem to account for.
B. The case with the first, more mechanical explanation is a bit more complicated.
  1. As we've seen, the basic laws of motion permit the collisions to "run backward."
  2. Still, if we can appeal to facts about collisions to provide a mechanism for entropy to increase, then we finally have a time asymmetry built into our system. Oddly, no mechanical account of how entropy is brought about has, to my knowledge, gained widespread assent.
  3. Thus, it's not clear that we have a mechanical explanation of the direction of time, and the more purely statistical explanation, as we've seen, would lead us to expect entropy to increase in both temporal directions.
C. The problem, then, is not merely that the laws of motion say that decreasing entropy should be possible but that it does not happen. Possible things fail to happen all the time. The problem is that thermodynamics, in particular the second law, seems to be in conflict with what the underlying laws of statistical mechanics would lead us to expect.
D. Here's a similar way of approaching the problem: If states of thermodynamic equilibrium are overwhelmingly the most probable ones, why is the world we observe so full of situations that are so far from equilibrium?
  1. Boltzmann suggests that we inhabit a peculiar corner of the universe where the thermodynamic equilibrium states that hold sway in most parts of the universe do not obtain.
  2. Only very peculiar combinations of circumstances will give rise to organisms that can think and observe. The reason we always see entropy increasing is because we inhabit a corner of the universe in which entropy is abnormally low. It has got nowhere to go but up.
  3. In other parts of the universe, entropy might decrease as much as it increases. Boltzmann suggests that this raises deep questions about the direction of time in those parts of the universe.

IV. The most influential answer to the puzzle about why entropy seems always to increase generalizes Boltzmann's suggestion beyond our corner of the universe. Entropy is on the rise everywhere because it started out very low everywhere.

   A. Even the mechanical explanation for the tendency of entropy to increase needs to posit a low-entropy state in the past. If entropy had started high, the mechanisms would help keep it there, but they wouldn't account for the overwhelming tendency for entropy to increase that we seem to observe.

   B. Thus, it seems that we must adopt the *Past Hypothesis*, according to which entropy started very, very (add about 10 to the 23rd power *very's*) low. If the universe is constantly moving toward more probable states, it must be moving from a *mighty* improbable state.

V. At this point, a major issue in the philosophy of explanation arises: Does the Past Hypothesis need to be explained?

   A. Here's a way of fleshing out the Past Hypothesis that makes it seem to cry out for explanation.

   1. Matter seems weirdly uniformly distributed around 100,000 years after the Big Bang. When dealing with an attractive force such as gravity, a uniform distribution of matter is highly unusual, because objects will tend to clump together.

   2. Huw Price, a philosopher of science, compares the Past Hypothesis to the idea of throwing trillions of foam pellets into a tornado and having them shake down into a uniform sheet, one pellet thick, over every square centimeter of Kansas. The Past Hypothesis differs from this mainly in being enormously less probable (according to some calculations, anyway).

   3. Furthermore, says Price, the Past Hypothesis is the only weird initial condition that we need in order to account for all of the low-entropy systems in the universe, because the improbable initial smoothness in the universe led to the formation of stars and galaxies, and these sorts of things are responsible for the temporarily asymmetric phenomena we encounter.

   4. Thus, deep facts about our universe seem to turn on an enormously improbable fact, namely, the incredibly low-

    entropy state of the universe at a certain point relatively soon after the Big Bang. Surely something that important and that improbable needs to be explained.
- **B.** But there are powerful reasons for wondering what could possibly explain such a fact and for wondering whether such an explanation would ultimately be scientific. Some serious empiricist worries loom large here.
    1. The Past Hypothesis can be compared to the *First Cause* argument for God's existence, and similar worries arise. Why carry the demand for explanation this far and no further?
    2. It will not help to explain a past state of surprisingly uniform distribution of matter by positing an even more improbable state before that.
    3. There's also a worry about initial conditions and single-case probabilities here. If universes were as plentiful as blackberries, we could pursue explanatory hypotheses about how they arise and develop.
- **C.** Does the Past Hypothesis count as a law? It is a prime example of something that happens only once that might still count as a law.
    1. It does not have the logical form we associate with laws. But it functions crucially in explanation of many different phenomena; thus, it might count as a law on a "systems" conception of laws, which identifies laws with axioms of true deductive systems that best combine strength and simplicity.
    2. Some think that calling the Past Hypothesis a law makes it stand in less need of explanation.

Essential Reading:

Price, "On the Origins of the Arrow of Time: Why There Is Still a Puzzle about the Low-Entropy Past," in Hitchcock, *Contemporary Debates in Philosophy of Science*, pp. 219–239.

Callendar, "There Is No Puzzle about the Low-Entropy Past," in Hitchcock, *Contemporary Debates in Philosophy of Science*, pp. 240–255.

Supplementary Reading:

Sklar, *Philosophy of Physics*, chapter 3.

Questions to Consider:
1. Does everything that is highly improbable call out for explanation? Why or why not? And which sense of probability (frequency or degree of belief, for example) figures in the notion of improbability at work here?
2. It is sometimes said that unique events cannot be explained, perhaps because explanation involves placing events in a pattern. But don't we sometimes explain unique occurrences? How do we do so?

Lecture Thirty-Three—Transcript
Entropy and Explanation

I'd like to begin winding down the course by looking at some examples from within the philosophy of particular sciences. There are a couple of reasons for doing this. For one thing, it's just plain useful to see a more detailed and, hence, more realistic treatment of how some of these general issues in the philosophy of science play out in closer contact with scientific practice.

In addition, most philosophy of science these days is philosophy of a particular science. This reflects the influence of Kuhn and of naturalized philosophy according to which it's important to pay serious attention to actual scientific practice because the "ises" of science determine, to a great extent, the "oughts" of science. Nevertheless, a kind of reciprocal illumination is also to be expected. We can understand some of the issues in general philosophy of science more clearly when we see them deployed less abstractly, with more factual content and with more context.

So, it's not the case that general philosophy of science simply gets applied unproblematically within particular sciences. The road between general philosophy of science and the philosophy of the more particular sciences runs both ways. New problems about explanation, about confirmation, about the meaning of scientific terms emerge from our reflection on particular sciences, not just on science as such. To some extent, applications sometimes do, and sometimes should, drive theory rather than theory dictating how the applications should go.

The illustration I've chosen for this lecture is from the philosophy of physics and will begin rather innocuously, but will end up raising some really quite surprising questions about explanation, about laws, and about scientific reduction—or so I hope.

The issue can be stated briefly, if rather peculiarly. Why can't I stir milk out of my coffee? This may not sound like the kind of "why" question that needs an answer, but it's less ridiculous than it sounds.

In one sense, it turns out I *can* stir milk out of my coffee (or I could if I actually put milk in my coffee). The basic laws of nature (both classical and quantum mechanical laws) permit milk to be stirred out of coffee. We will assume a nice, simple Newtonian mechanics that will do no harm in our context. The basic laws of motion permit all the gas in a container to cluster in one corner of the container, and they permit heat to flow from a metal bar that's been kept in the freezer to a metal bar that's been kept in a hot often. There's nothing in the basic laws of motion specifying in which direction molecules are supposed to move.

For any actual motion of molecules, a reverse motion is permitted by the basic mechanics. Just change all the velocities that are positive to negative and all the velocities that are negative to positive. You've violated no law of nature; you get a physically permissible motion.

But though those things are possible, they never happen. They're permitted by the basic laws, in this case roughly statistical mechanics (the science of molecules in motion), but the laws that in some sense reduce to (or at least are thought to reduce to these more basic laws), the laws of thermodynamics, tell us that air never leaks into a punctured tire, and a hot bowl of soup does not spontaneously heat up when brought to the dinner table. So, the underlying laws tell us that these things can happen, while the overlaying laws tell us that they never do happen.

The second law of thermodynamics (as we saw some lectures ago, thermodynamics concerns such phenomena—directly observable phenomena in a loose sense—as temperature and volume; it's a phenomenological science) says, speaking loosely, that energy tends to spread out, to dissipate. Like some of my graduating seniors, concentrations of energy become dissipated and increasingly unavailable for work. That's the second law of thermodynamics—again, loosely stated.

Another way of saying this is that *entropy* (this is a term that will figure frequently in this lecture; it is a measure of this dissipative tendency) will tend to increase. We don't need a rigorous understanding of it; we just need a basic understanding. A system that is energetically isolated (a system from which energy is neither added nor removed) will tend to move toward an equilibrium state, one that is unchanging at the phenomenological or macroscopic level (in other words, something like heat will spread out, and it will stay spread out).

The question is, why does the second law hold, since there's nothing about the underlying laws of motion—taken just by themselves—to make the second law hold. This question started to receive a lot of attention in the middle of the 19th century, mainly occasioned by the steam engine, which is about turning heat into work, and things like that.

Starting in the 1870s, Ludwig Boltzmann, an Austrian physicist and mathematician, worked out the two most influential answers to this question of why the second law of thermodynamics holds, though it seems to have taken him a while to appreciate that he had worked out two quite distinct answers to the question.

Boltzmann's first answer is that the effect of collisions between rapidly moving gas molecules will tend to bring it about that entropy will increase until it reaches a maximum value. It was soon pointed out, though, that since the underlying laws of mechanics permit motions in the opposite direction of the motions that actually happen, this explanation will be statistical, not deterministic. This touches on some issues we've raised earlier in this course. Well before the rise of quantum mechanics, we have the idea broached that the second law of thermodynamics is probabilistic, not deterministic, and that this is a basic law. There are possible motions by which entropy would decrease, even though it seems always to increase.

Boltzmann's second answer builds on this realization about his first answer. Rather than claiming that collisions among molecules cause entropy to increase, he suggests in the second answer that there are just more ways for particles to be spread out than there are for particles to be concentrated. It's a matter of different degrees of what we have called *multiple realizability* in our discussion of scientific reduction.

So, the idea here is that you don't need an interesting causal story about why you generally get dealt a bad hand at poker. Assuming a random distribution of cards, there are just a lot more bad hands than there are hands that make, say, a straight flush. Given just random changes in motion, running the analog here, if an isolated state is in a low-entropy state—a not very dissipated state—it's highly likely to move to a higher-entropy state simply because there are so many more higher-entropy states than lower-entropy states.

So, on this interpretation, the second law, again, becomes statistical, and it stems from the fact that almost all of the possible distributions of particles at the micro-level, together, realize high-entropy states, high-dissipation states, at the macro-level.

So far, so good. Isolated systems tend toward equilibrium states either because of the laws of mechanics, or because most of the possible states are themselves equilibrium states.

The first explanation has the second law of thermodynamics serve as a causal law in a pretty robust sense: Collisions *cause* increases of entropy; there is a real tendency in nature that operates to bring about increasing dissipation. This is most naturally construed as a version of the necessitarian conception of a law that we saw back in Lecture Twenty-Two—it makes entropy increase.

The other explanation looks friendlier to an empiricist or a regularity conception of law. On this view, the tendency towards entropy is just the sort of thing that always happens. Nothing is posited to *make* it happen. The tendency towards entropy is just what's to be expected, given the laws of nature—but there's no making, there's no mechanism invoked.

Either way, though, we're going to run into a peculiar puzzle. Let's take the more empiricist explanation first. This explanation of why entropy increases into the future should equally well explain why it increases into the past. That's a weird formulation; let me unpack that. Just as almost all of the states a closed system can move to are high-entropy states (because of the way that probabilistic distributions work; the explanation here is more or less mathematical), but the same explanation indicates that just about all the states it could have moved from are high-entropy states. So, the same explanation should work in both temporal directions.

For the same reason that non-equilibrium states move towards higher entropy, they should themselves have come from higher-entropy states because there are so many more higher-entropy states than there are lower-entropy states. So, entropy should increase as we move towards the past, just as it does as we move towards the future. But that never happens. That would be things concentrating as they move towards the present. There's a *temporal asymmetry* at the observable, thermodynamic level that this explanation doesn't seem to account for.

The other—the more causal explanation—might seem more promising in this respect. There's still a worry about a time asymmetry here because—though I didn't put it this way initially—our earlier point about the physical possibility of the motion of particles being reversed shows that the basic laws of motion are *time-symmetric*. Every motion that can happen forward in time can happen backwards in time. You can, as it were, run the universe's movie backwards, without violating any laws of nature. This is a kind-of peculiarly under-appreciated fact about the laws of mechanics.

Nevertheless, if we can appeal to facts about collisions to provide a mechanism for entropy to increase, then we finally have a time asymmetry built into our system. It ends up, though, being very complicated trying to formulate what molecules and their collisions must be like in order for these collisions to bring about increased entropy. There are other mechanisms (for instance, from quantum mechanics) that have been proposed, but there is no generally accepted mechanism that would provide this time asymmetry.

This is remarkable because we're looking for a law on which to base the direction of time. Without such a law, without such a causal mechanism, we don't seem to have any reason for thinking that heat, gases, that sort of stuff, should be moving in the direction of dissipation rather than concentration. If we don't have a mechanism to account for it, and if the probabilistic distributions seem neutral between those directions, we have a problem because the laws of motion don't have a built-in temporal direction. So, this would lead one to expect that entropy would be increasing in the direction of the past, just as it seems, in fact, to increase in the direction of the future.

The problem is not merely that the laws of nature say that decreasing entropy (in a closed system, to be fancy about it) should be possible, and that this possible thing doesn't happen. That needn't be a problem at all. All sorts of things that the laws of nature say are possible fail to happen. That's not an objection to a theory. The point is that thermodynamics—in particular, the second law of thermodynamics—seems to be in conflict with what the underlying laws of statistical mechanics would lead us to expect to actually happen, not just to be possible.

Here's another way of putting this problem: If the thermodynamic equilibrium states are overwhelmingly the most probable ones, why is the world we observe so full of situations that are so far from equilibrium? If those are the dominant probabilistically favored states, why don't they happen more often? Boltzmann, in the 1890s—about 20 years after first raising these issues—suggested that we inhabit a very peculiar corner of the universe, where the thermodynamic equilibrium states that, in fact, hold sway in most parts of the universe do not actually hold. So, we're just in a weird neighborhood from the standpoint of the universe.

Why would that be? Well, Boltzmann has something of an explanation: Only very peculiar combinations of circumstances will give rise to organisms that can think and observe. Critters like us require concentrations of energy and other conditions that are associated with low-entropy states. So, the reason that we happen to see entropy increasing all the time is because we inhabit a corner of the universe in which entropy is abnormally low, and so the main direction for it to go is up. So, it's not a universal truth; it's a local truth.

Furthermore, what we mean by the future is tied to the direction of local entropy increase. In some parts of the universe, on Boltzmann's view, there is no objective distinction between past and future because there is no direction of entropy increase—because things are at something like an equilibrium, so there's no direction of dissipation or concentration.

We're getting into some pretty deep waters here, but think about a part of space where there is no local gravitational field acting—just far enough away from all objects that there's no gravitational pull. Which direction is down? That's a question that doesn't have a good answer. Boltzmann thinks the same thing can be said about past and future, in big chunks of the universe—there's just no answer to what counts as past and future in those places. He even says that the direction of time that counts as future, in some parts of the universe, could be opposite to the direction of time that counts as the future in our corner of the universe. Future could run in the direction of concentration in parts of the universe.

There turn out to be some problems with Boltzmann's explanation of temporal asymmetry as a kind of local phenomenon. We have other fish to fry; we won't be able to go into those, highly interesting though they are.

The most influential answer to this overall time asymmetry problem is actually a kind of generalization of Boltzmann's idea that locally, we see entropy on the rise because—around here anyway—it started so low in the first place and doesn't have, really, any place else to go.

Many physicists and philosophers posit a very low-entropy starting point for the entire universe. On this view, nothing "makes" entropy increase; there is no deterministic basis for the second law of thermodynamics. It's just overwhelmingly likely, given the starting point.

It's worth noting that a posit like that seems to be required—not just according to Boltzmann's purely statistical explanation of the second law, but also according to his causal explanation of increasing entropy. Why? Because if entropy had started high—as it in some sense it should, given the number of possible high-entropy states compared to the number of possible low-entropy states—then the mechanical explanation would only help keep it there—it won't bring it about. So, whether we appeal to an entropy-increasing mechanism or not, we need entropy to have started low for it to be able to increase this much. When we say we need entropy to have started low, we mean really low.

That means—according to some standard interpretations of this—that the starting point will have to have been very, very, very (add a few thousand "verys") improbable. One calculation has it that the entropy starting point of the universe had a 1 chance out of 10 to the 10 to the 23rd power of being chosen randomly (according to the measure that's standard in statistical mechanics). It's hard to explain that in more detail; the point is just the low-entropy starting point of the universe is immensely improbable on standard measures.

Why? If the universe is pretty much always moving toward more probable states, then the initial state will need to have been mighty improbable. That's the basic point. There's some interesting work here to be done about the various senses of "probability" that figure in this. If you're not careful, you'll move from one of the interpretations of probability we've talked about (probabilities as frequencies) to probabilities as degrees of belief. We won't go into enough detail to have to belabor this point.

But now some pretty deep questions about explanation, evidence, and the business of science start to loom. To what extent does the Past Hypothesis (which is the name usually given to this claim about the low-entropy origin of the universe) call out for explanation?

Here's one way of understanding what the Past Hypothesis says. Matter seems to have been weirdly, uniformly distributed about 100,000 years after the Big Bang—which is no time at all, cosmologically speaking. A uniform distribution sounds like a high-entropy state to us; it sounds like a dissipated state. But that's because we're used to thinking of entropy in terms of things like gas molecules, and the main forces there are pressure; that's a repulsive force.

When you're dealing with objects that are primarily governed by an attractive force like gravity, a uniform distribution of matter is highly unusual because it's highly unstable—matter will tend to clump together. So, this is the surprising fact—the uniformity of the distribution of matter in the universe relatively soon after the Big Bang.

Huw Price, a philosopher of science who does a lot of work on the direction of time, compares the Past Hypothesis to the idea of throwing trillions of foam pellets into a tornado, having them shake down into a uniform sheet—one pellet thick—over every square centimeter of Kansas. According to the standard measures (there are some quibbles here about exactly how they should be run), the Past Hypothesis is immensely more unlikely than the tornado hypothesis.

So, says Price, the Past Hypothesis is a very weird initial condition, and it turns out to be the only weird initial condition that we need in order to account for all of the low entropy systems in the universe. This weird initial smoothness leads to the formation of stars and galaxies—given some nice, well-behaved laws—and it's these sorts of things (stars and galaxies) that are responsible for the temporally asymmetric phenomena we encounter.

So, the deep facts about our universe seem to turn on a distinctively and enormously improbable fact—namely, the incredibly low entropy state of the universe at a certain point relatively soon after the Big Bang. And in some sense, given the laws, they turn only on that fact. So, some philosophers want to say, what would call out for explanation more than such a jaw-droppingly surprising and important fact? So, it seems obvious—given those considerations—that an explanation of why the Past Hypothesis is true (assuming it is) would be a prime task for physics and for cosmology.

On the other hand, however, there are powerful reasons for wondering what could possibly explain such a fact, and for wondering whether such an explanation could ultimately be scientific.

The worries here are broadly empiricist in flavor. The Past Hypothesis can be compared to the God Hypothesis as put forward in what is usually called the *cosmological or first-cause argument for God's existence*. This argument was made famous primarily by Thomas Aquinas. Oversimplifying somewhat brazenly, but I hope at least a bit helpfully, the argument says everything needs an explanation, and so the universe needs an explanation, and so this explanation is God. That is far from a full-dressed presentation of the argument, but you get the idea.

That puts you in a position to appreciate the basic worries that arise (and I don't, by any means, mean to suggest that a smart guy like Aquinas wouldn't have some replies to these worries). The major worry is why do you stop at God? If absolutely everything needs a cause, needs an explanation, then it looks like God should need one, too. If, choosing the other horn of the dilemma, some things can be self-explanatory—or can be posited as just brute facts ("stuff happens"), why can't the universe itself or the Big Bang get that status that is reserved for God? I don't mean there's no answer to these questions from the standpoint of the cosmological argument; I just mean they're serious questions.

The point is that empiricists are suspicious of the demand for explanation being pushed too far. When we start to run out of possibilities for getting evidence that bears reasonably directly on a hypothesis, the project of explanation (if one is an empiricist) needs to stop.

So, the parallel worry about the Past Hypothesis goes something like this: It's not going to help to explain a past state of a surprisingly uniform distribution of matter by positing an even more improbable and surprisingly uniform distribution of matter before that. That would be a good explanation of the state, since the tendency for states is to move towards equilibrium, and so a more improbable state will explain the slightly less improbable state. But there's no overall explanatory gain, since the explaining state would seem to be even more in need of explanation than the explained state was.

So, the empiricist says, why not stop before we start? Just admit that our universe started from some unbelievably improbable state. There's a sense in which most things that happen are pretty improbable, so it's not obviously a demerit of our theory if we have to admit that the universe began from an improbable state.

Another way of pressing this point is to say that the Past Hypothesis is more or less an initial condition (not literally an initial condition—it's after the Big Bang—but it's close to an initial condition) of the only universe we know. As Charles Peirce (the philosopher on whom I wrote my dissertation, an American pragmatist) said, if universes were as plentiful as blackberries, we'd be able to study them. We could run some tests and we could get observation to bear on questions like how these critters called "universes" arise and develop.

This is the problem of single-case probabilities. It's hard to place the origin of the universe in a reference class from which we can draw samples and get data about how these things work. So, the empiricist says, without being able to do that, it's not clear how we're going to get evidence to bear on the question of how these things get the initial states they get. We can't run tests; we can't look at a sample.

Empiricists also tend to mistrust our intuitions about what calls out for explanation and why. The classic case here is Newton on gravity. Newtonian gravity seemed, to his critics, especially the Cartesians, to need a mechanical explanation, some kind of contact force, a push or a pull that can't happen instantaneously at a distance. But attempts to explain how there could be one ended up positing untestable metaphysical objects and properties (aethers and things like that pushing objects around, but that couldn't independently get detected).

So, for centuries, until Einstein came along, physicists learned to accept gravity as a brute fact not intrinsically in need of explanation; it's just a basic piece of the furniture of the universe. The intuition that contact forces don't require explanation, but that forces that act at a distance do require explanation, had been powerful, but it got rejected for centuries (Einstein arguably is a kind of reversion to a more Cartesian or Aristotelian picture, but let's put that aside for right now).

Another interesting issue that arises in this context concerns whether the Past Hypothesis counts as a law of nature or not. It's a prime example of something that only happens once, but that nevertheless might count as a law. It's both an initial condition and law-like in an important respect (this is going back to Lectures Twenty-One and Twenty-Two, for those of you with good memories). It doesn't have the logical form we associate with laws of nature; it's not of universal conditional form or "all A's are B's." But it functions crucially in explanations of lots of different phenomena, and so might, for instance, on a "best systems" conception of laws—which identifies laws with the axioms of the true deductive systems that best combine strength and simplicity (maybe that sounds familiar). The Past Hypothesis might well be an axiom. Why? Because when we add it, we can derive lots and lots of correct predictions that we couldn't have derived without it.

Generally, a statement is going to be law-like in order to do that. You won't be able to plug in a particular fact and get all of that sort-of observation-deriving juice out of it. But the Past Hypothesis is special; it's a particular fact that is very explanatorily rich.

Does calling the Past Hypothesis a law, rather than a fact, make it any less needful of being explained? Some think so, because the laws determine what counts as physically possible in our universe. So, on a view like that, we can say it's physically impossible for entropy to have been higher than it was in the past. Why? The laws of nature include the Past Hypothesis—so given the laws of nature, the Past Hypothesis not holding is impossible.

That's certainly not to explain the Past Hypothesis, but some philosophers think it's a legitimate reason to reject the demand that the Past Hypothesis get explained.

We have only scratched the surface of this debate; I don't know enough physics to dig too deeply into this debate. Much is going to turn, at least for somebody who's sympathetic towards philosophical naturalism, but arguably for almost anybody, on what room our current best science leaves for possible explanations here. What might—according to science (not some philosophical theory of explanation)—count as a possible explanation of the Past

Hypothesis? That's going to tell you a lot about the extent to which you think it should be explained.

We don't want to assume naturalism to be true. One might—with some plausibility, I think—claim that abandoning the search for an explanation of such a remarkable fact (or law) could constitute a betrayal of scientific and intellectual standards no matter whether our current science says anything about how to explain this hypothesis or not. You might be willing to defend a kind-of a priori demand that amazing and important stuff should receive an explanation, even if we have no idea how to provide one. That's to reject the kind of empiricist constraints that the demand for explanation tends to push towards metaphysics, and metaphysics is a temptation that should, by and large, be resisted.

So, this debate between empiricists and realists about the conditions under which explanatory inferences are possible and the conditions under which they're desirable comes to a head rather nicely around this issue. And I don't think there's an easy answer. So, one question is: How bad is it to leave astonishing facts unexplained? On the other hand, how bad is it to posit untestable explanations for astonishing facts? Each of these seems to run afoul of something deep within our scientific norms. So, the case illustrates rather nicely a pretty profound tension between the evidential boundaries that seem built into science and the explanatory ambitions that seem equally built into science.

Next time, we'll continue illustrating the philosophical questions that arise within particular sciences by looking at the status of species within biology: Is this a real classification system, or a convenient classification system, or is there no difference between those two questions?

Lecture Thirty-Four
Species and Reality

Scope:

What kind of a kind is a species, if indeed it is a kind at all? We certainly talk as if species have properties of their own (such as being endangered), and in fact, a species is, in many respects, more like an individual than it is like a class or kind. But biology defines species in a number of ways, and even some of the best definitions seem to exclude most organisms on Earth from being members of a species. In this lecture, we try to understand the motivations behind biological classification, and we wonder about the things so classified. How are we to decide whether a species concept is a good one? And how are we to decide whether a good species concept tracks something real?

Outline

I. We now turn to the philosophy of biology, probably the most rapidly expanding field within the philosophy of science. We will focus on the notion of a species and use that as a window into parts of the philosophy of biology and into general philosophy of science issues about classification and reality of scientific kinds or classes.

   A. The species concept figures centrally in biology.
   1. Species are fundamental units in the story of evolution. They are born, split into new species, and eventually become extinct.
   2. They are also fundamental to biological classification. Members of the same species have something biologically important in common.
   3. As uncontroversial as these claims sound, together, they put real pressure on the notion of a species. It is not obvious that the notion can serve both these functions well. The properties of organisms and populations that are relevant to the story of evolution might be different from those that are important for certain classificatory purposes.

      4. Furthermore, species loom large in our applications of biological thinking, for example, in some environmental protection laws.
- **B.** The very notion of a species can be made to seem puzzling.
    1. Evolutionary change is, in some important sense, gradual; new kinds of creatures arise via small mutations in existing creatures. Where, then, are we to find distinctions of kind within a fundamentally continuous process?
    2. Just as there is impressive continuity across species, there is impressive variety within species. *Conspecifics* (members of the same species) are not united by a common essence. No genetic, phenotypic, or behavioral trait is essential to making something a member of a species. Nor are there, in general, traits that are unique to a given species.

II. This raises the issue of the ontological status of species: what kind of entity is a species? One surprising but common answer is that species are more like individuals than like classes of objects.
- **A.** A kind of spatiotemporal or causal connectedness is required for a species. If evolutionary processes are primary, it is plausible to hold that something has to be part of a lineage to be part of a species. And a lineage is a particular thing, not an open-ended class of things.
- **B.** On this view, species have a beginning and an end in time, and they have a spatial location.
- **C.** Perhaps most importantly, species are constituted by properties at the population level, not at the level of the individual organism. This gives a species a certain cohesiveness needed to play a role in scientific explanations. It is the population, not individual organisms, that bears such properties as "having lost much of its genetic diversity" or "having a trait that was rare become prominent." These are taken to be genuine and explanatorily important biological properties.
- **D.** These population-level properties can change rapidly, in accordance with our sense that boundaries between species are relatively stark. This is part of the explanation of how

gradualism and continuity at the level of parent-offspring genetic relationships can be reconciled with the idea that organisms seem to come in distinct kinds.

III. A surprising number of definitions of *species* have been proposed by biologists.
- **A.** *Phenetic species concepts* group organisms in terms of genetic, behavioral, or morphological similarity.
    1. But similarity, as we've seen in this course, is a tricky notion. How is it to be measured, for instance?
    2. Intraspecific similarity is less impressive than one might think. Queen and drone bees don't appear all that similar.
    3. If the notion of a species is determined by similarity, then species membership can't be used to explain similarity.
- **B.** According to the *biological species concept*, a species is a group of organisms that can interbreed and produce fertile offspring. A species is characterized by the relatively free flow of genes within it.
    1. This concept is difficult to apply over time. Let A, B, and C be members of successive generations. Suppose that B is about equally similar to A and to C and could breed with either one but that A and C would not be able to breed. If A is the standard, then C is a member of a new species, but if B is the standard, there is just one species. Proponents of this species concept apply it only at a time, not over time. But that means they need an independent notion of a speciation event in order to determine whether a creature is conspecific with a given ancestor.
    2. The notion of a reproductively isolated population is also tricky. Not all kinds of reproductive isolation (such as being kept in a zoo) count.
    3. The most glaring problem with the biological species concept is that, as it stands, it does not even apply to creatures that reproduce asexually. Furthermore, gene flow is much easier among plants and single-celled organisms than it is in multicellular animals.

- C. *Phylogenetic species concepts* define species in terms of a shared history. The biological concept involves a theory about the process whereby species are created and sustained, while phylogenetic accounts simply appeal to patterns of common ancestry.
  1. They thus make room for the idea that mechanisms other than reproductive isolation can produce speciation.
  2. So far, this is just a *grouping criterion*: It lumps organisms together in ways that matter to evolution, but it does not provide *ranking criteria*; it doesn't tell us which groups are species or genera or sub-species and so on.
- D. The *ecological species concept* identifies a species as a group of organisms sharing a particular ecological niche.
  1. This looks like an attractive way to handle asexually reproducing species, because they do not compete with one another for mating opportunities, but they do for roles within the ecosystem.
  2. But how well do we understand ecological niches and how enduring are they?

IV. How important is it to unify these various conceptions of species?
- A. Monists think there is a single correct species concept, but monism runs the risk of excluding species concepts with genuine explanatory power. Might multiple ways of identifying species each answer to legitimate scientific purposes?
- B. Pluralists think that there is no problem having multiple conceptions of species. Pluralism comes in degrees, but the more tolerant one is of different species concepts, the more an explanation seems to be needed of what makes each of them a *species* concept.
- C. Some thinkers are skeptics about species. They deny that anything in the world answers to all the uses to which the notion of species gets put.

V. These issues about the reality of a grouping or category arise even more clearly in discussions of higher taxa, such as families, groups, and genera. As with species, pluralists will stress the legitimacy of different purposes served by classification, while monists will point out conceptual problems that seem to stand in the way of thinking of different classifications as right or real.

A. A *phenetic* classification system would try to convey information about similarity in a maximally efficient way. Not only are species maximally similar organisms, but genera are maximally similar species, families are maximally similar genera, and so on.

B. A related but distinct approach would be to classify in terms of *evolutionary disparity*. This is simultaneously a historical and morphological system. This approach might accord a lizard species sufficiently different from all other species of its own genus.

C. Finally, we might classify in a way that reveals evolutionary history—organisms classified in terms of ancestry. This is the *cladistic* approach to classification.

D. Pluralists might be tempted to allow all three kinds of classification.

E. Monists (and others) might object to the phenetic and evolutionary disparity systems on the grounds of the unclarity of similarity relation. They might also object that the cladistic approach can't make room for groups not united by a common ancestry but nevertheless partaking of a genuine explanatory role in evolution.

F. Cladism is easily the most popular approach to classification, and it has a theory of which biological groups are real—those that share an ancestral species. A reptile is not a real category for cladists because there is no species that is ancestral to all reptiles that is not also ancestral to birds. But this notion of real groups does accord any status to such levels as genera and families. All groups that share a common ancestor species are legitimate and all that do not are not legitimate.

G. Cladists, monists, pluralists, skeptics, and others all appeal to implicit notions of what makes a group or a distinction real. This discussion helps flesh out our distinction between hard and soft realisms back in Lecture Twenty-Six.

Essential Reading:

Sterelny and Griffiths, *Sex and Death: An Introduction to Philosophy of Biology*, chapter 9.

Supplementary Reading:

Sober, *Philosophy of Biology*, chapter 6.

Questions to Consider:

1. Would you value a distinctive group of animals less if you were to be convinced that it constituted a sub-population but not a species? How, if at all, do various species concepts hook up to what we *value* about species?
2. To a committed defender of a phylogenetic species concept, there is no such thing as a reptile, because the animals we call reptiles don't share a distinctive common ancestor. Does this convince you that the category "reptiles" is illegitimate? Why or why not?

Lecture Thirty-Four—Transcript
Species and Reality

Last time out, we got a small taste of what the philosophy of physics looks like—and also what issues about laws of nature, explanation, and reduction look like from a standpoint a bit closer to ongoing scientific work than that provided by our general survey of philosophy of science. We could, of course, have pursued a number of equally fascinating issues within the philosophy of physics. There's the issue about what the correct geometry of space is; whether and in what sense space-time is a kind of object, thing, or stuff—rather than a set of relations; and a host of problems that arise within quantum mechanics.

We turn now to the philosophy of biology, probably the most rapidly expanding part of the philosophy of science. Again, we find a kind-of embarrassment of riches. There are questions about what the units of selection are. Is it organisms, or genes, or species that get selected among, or sometimes do each of these categories figure in selection explanations? Do genes code for phenotypic traits, for traits at the level of observation, or is the relationship between genes, and proteins, and traits more complicated than would allow us to say something as straightforward as the idea that a gene codes for a trait? We don't talk about the environment coding for a trait, even though it helps determine what traits an organism has.

There are issues about to what extent our biological nature helps explain human behavior. And there's the issue about the extent to which we have a conceptual grip on the notion of what it is to be alive or to be a living thing. This is just a sample, of course, and not all of those issues lend themselves equally well to a half-hour lecture.

So, we're going to focus our attention in this lecture on the notion of a species, and we'll use that as a window into several aspects of the philosophy of biology and into general issues in the philosophy of science about classification and the ways in which classification tries to hold itself responsible to reality. What is it for a classification to be legitimate or correct, or to be scientifically applicable, and what does that show us about the things classified in that way?

The species concept figures centrally in our thinking about biology. That sounds like an empty truism, but it's a bit more interesting than it seems. Because, as we'll see, many people think that the higher-level biological classes (sometimes called *higher taxa*—genus, family, kingdom, that sort of thing) don't have any particular role in biological thinking. So, the idea that the species is the classification that's of biological importance is noteworthy.

What kind of importance do species have? For one thing, they are fundamental units in the story of evolution; species are the actors in evolution. Evolution is not especially concerned with individual organisms; it's species that are born, that split into new species, and eventually become extinct.

Species are also fundamental to biological classification. Species are supposed to have something biologically important in common.

Again, this sounds like a truism, but there's more going on here than meets the ear. It's not obvious that the notion of species can serve these two masters equally well. There's an assumption that whatever figures in the story of evolution will also figure in classification, but it's not obvious that things are going to clean up as nicely as that. Evolutionary biologists want to understand how certain things happened, how a certain process works, how organisms came to be the way they are. For those purposes, it's okay if the notion of species ends up being pretty messy, if rather different mechanisms can bring organisms together into groups that matter for this story—the groups that undergo speciation, extinction, and things like that.

But a messy, multifaceted species category is going to be a lot less useful for classifying purposes than it might be for purposes of explaining the history of organisms. Biologists who are in the classifying business are interested in conveying information. They need relatively enduring structures so that the classification tells you what to look for if you want to identify a bird species, or what it's going to look like if you dissect a certain kind of creature. And that's a different task than trying to accurately relate the twists and turns of a particular story (namely the story of evolution).

So, while there's presumably some kind of connection between the units that undergo various evolutionary events (like speciation) and those that can be identified and classified by examining the organisms themselves, rather than their historical properties, we've already managed to commit some philosophy with the idea of a species. We've taken a category that we would have thought was pretty simple and natural, and we've made it look unfamiliar, and puzzling, and almost two-headed because it's got to serve these classifying and these historical interests.

As if the job of being a species concept is not complicated enough, species loom quite large in our applications of biological thinking (for instance, in our environmental protection laws). So, if the category of species turns out to be problematic, that raises at least some initial questions about the use we can make of this concept. Why, if at all, are species—rather than individual organisms, populations, genes, or what have you—the things at which preservation efforts should be directed?

I don't mean to suggest that there's no good answer to that question. I mean to suggest that the question is, perhaps, more interesting to ask than we might have thought. And there are additional reasons why reflection makes the species category seem increasingly puzzling.

For one thing, evolutionary change is—in an important sense—gradual; new kinds of creatures arise via really quite small mutations in existing creatures. Massive mutations of a sort that would clearly and decisively distinguish offspring from a parent are highly unlikely to result in viable offspring. Huge mutations just don't lead to creatures that can survive and reproduce.

So, where are we going to find the distinctions of kind within a process that looks fundamentally continuous? Why is it that we find different kinds of creatures rather than a continuous, as it were, spectrum of creatures within, say, the primates?

And just as there is impressive continuity *across* species, there is impressive variety *within* species. Conspecifics (that is, members of the same species) need not resemble each other in any clear or decisive ways. There are butterfly species, the members of which mimic different other butterfly species—so each of two butterflies from the mimicking species might resemble butterflies of other species more closely than it resembles members of its own species.

And the philosophers like to use chemical elements and biological species as our clear, classic examples of natural kinds. It's far from clear that species work very much like the chemical elements do. There's nothing in a species that seems to answer to an atomic number that makes the element the element it is. There's no genetic, there's no phenotypic, there's no behavioral trait essential to making something a member of the species of which it's a member.

Why not? There are all of these forces that tend to lead to biological change (things like mutation and genetic drift), and they're going to tend to make it the case that there's always a reasonable possibility—if not yet in actuality—that some member of a species could lose any particular trait (through, say, a minor mutation) and it wouldn't thereby forfeit its membership in the species.

Nor are there, in general, traits that are unique to a given species. This is different from the case of chemical elements. Not only is it the case that all samples of pure gold share an atomic number—it's also the case that no other substance shares gold's atomic number. We've just seen that members of a species needn't share any particular trait, and it's also true that even if they do, there's not much that prevents some members of another species from also sharing that trait.

So, one thing species don't seem to be is one of the things they're classically taken to be—namely, natural kinds "out there" united by a common and a distinctive essence. Because that would require that the species have a property that is unique to it and essential to it, and that seems to sit badly with the biological facts.

There are less demanding conceptions of biological kinds that species could be taken to fit. These are kinds united not by an essence, but by stable clusters of certain properties that work together in ways that will get too complicated for us to go into. This is a reasonably popular view with those who want to put the notion of a species primarily to classifying use, rather than to the purposes of understanding evolution, which is not the main use to which the species concept gets put, as we'll see. Even for classifying purposes, it's not clear that a view like this, though it's much more permissive than the essential property view, this view still might not be able to do justice to the range of traits and trait-clusters that can occur within a species.

All this raises the issue of the *ontological status* of species. If a species is not a natural kind in that full-blooded sense—a group of organisms united by an essence—what is a species, and what makes individual organisms part of a species?

One answer to this question—more common among those who want to put the notion of a species to evolutionary rather than taxonomic use—is really quite surprising. The idea is that species are more like individuals than they are like classes of objects.

Why is that? They require that a kind of spatiotemporal or causal connectedness constitute something as a species. If an identical copy of a Bengal tiger arose on a distant planet (somehow, miraculously) and were brought here, it arguably would not be a part of the terrestrial species. If the evolutionary processes are central to what makes something a species, it's plausible to hold that something has to be part of a lineage in order to belong to a species. And a lineage is something like a thing rather than a class of things. A lineage is something that is casually connected.

So, on this view, it's literally true that species have a beginning and an end in time. It's all but literally true—if not literally true—that species are born and die. They don't quite have to be organisms, but they are individual things on this view. Species have a spatial location, a range, though that need not be spatiotemporally continuous.

Perhaps most importantly, species are constituted by properties at the population level, not at the level of the individual organism. Let me explain; this is a crucial point because this gives species a kind of reality, a kind of cohesiveness. A species is a *unit* of evolutionary change.

Because it's the population, not individual organisms, that bears many of the properties that matter to evolution. The population can literally have a property like "having recently lost much of its genetic diversity" or "having a trait that was once rare become prominent" within the population. If these are genuine and explanatorily important biological properties, they are borne by populations, not by individual organisms.

These population-level properties can change quite rapidly, and that's the explanation for our sense that the boundaries between species are relatively stark, even though they arise via very slow, continuous processes of things like mutation. This is how we can reconcile that gradualist picture—that picture of continuity at the level of parent-offspring biological relationships while also making sense of the appearance, anyway, that organisms seem to come in kinds rather than in a continuous spectrum; there's no living missing link between humans and other primates.

So, for instance, when a population encounters a serious new disease, there won't be much change within individual organisms from before the disease arrived to after. But the population will very quickly change. It will become smaller, and it will equally quickly become characterized by having a much higher percentage of its members resistant to that disease. So, the population changes starkly.

Insofar as species are more like individuals than they are like classes of things, an individual organism is more literally a part of a species than a member of it, in the way that my finger is a part of me. So, the idea is that a lineage is more or less an individual thing, and so it has components, not instances or members. So, no similarity of form, no similarity of DNA or behavior can do the trick of making something a part of a lineage; that's got to be a matter of a kind of causal connection, a place and a story.

This brings us directly to the question of how we should best define species. A surprising number of definitions of "species" have been proposed and are taken seriously by biologists."

Though they were once common and are less common these days, *phenetic species concepts* are a good place to start. They group organisms in terms of similarity. It can be genetic similarity, behavioral similarity, or morphological (that is, bodily) similarity.

Similarity, as we've seen over and over in this course, is a tricky notion. There are too many ways, for instance, that emeralds can be similar in color to the emeralds we've already observed. They can share the color "grue," or the color "green," or the color "gred," and this is the problem that faces phenetic species concepts.

How is similarity to be measured, and on what facts is it to be grounded? To what extent are we entitled to think of ourselves as tracking distinctions that are "out there" rather than what we happen to care about when species membership is to be determined by a similarity measure?

Intraspecific similarity, similarity within a species, as we've noted, is a good bit less impressive than one might think—especially if you don't help yourself to the assumption that members of a species must be deeply similar. If you just look at queen and drone bees, they're members of the same species, but they don't look or act all that alike. And in many species, males and females aren't all that similar.

Furthermore, if the notion of a species gets determined by similarity, then species can't get used to explain similarity—but it seems that we'd sometimes, at least, like to explain similarity in terms of membership in the same species. So, the once-popular phenetic species concept is in a bit of trouble.

Much more popular these days is the *biological species concept*, according to which a species is a group of organisms that can interbreed and produce fertile offspring. Species on this view are characterized by the free flow of genes within a population.

This is probably the dominant species concept, but it has some problems. As it stands, it's difficult to apply over time. Let A, B, and C be members of successive generations. Suppose that B is about equally similar to A and to C. So, B could breed with either A or C, but A and C would not recognize each other as possible breeding partners. So, if A is the standard, then C is a member of a new species (since there's no possibility of gene exchange), but if B is the standard, there's just one species. It starts to look arbitrary how we divide a lineage up over time.

For reasons like this, proponents of the biological species concept apply it only *at* a time, not *over* time. But that means that in order to distinguish a species from just a lineage, however large, they need an independent notion of a speciation event in order to determine whether a creature is of the same species as a given ancestor. This is provided by a notion of reproductive isolation. It's a story about how species are born and how they preserve their distinctiveness when a gene pool gets sufficiently isolated such that genetic innovations can be preserved.

But this notion of reproductive isolation is a little bit tricky. Reproductive isolation comes in degrees, and not all kinds of isolation (for instance, being kept in a zoo) should count. Zoo animals are reproductively isolated, but that doesn't make them their own species. Furthermore, there are ring species, which are characterized by each population being able to breed with nearby populations, but unable to breed with more distant ones. So, that's got that transitivity problem that I was presenting as a problem for this species notion over time. That can happen *at* a time with a ring species.

To make a more purely philosophical point, modal notions—like necessity and possibility—generally strike philosophers as problematic, and so we'd like to hear more about the sense in which populations are members of different species if they cannot interbreed. A word like "cannot" always jumps into a philosopher's ear. Empiricists, for instance, will want to know how claiming that two organisms "cannot interbreed" is distinguished evidentially from just saying that they "do not interbreed."

The most glaring problem with the biological species concept, at least as it stands, is that it can't apply to creatures that reproduce asexually—which is arguably the most common kind of reproduction among species on Earth, because it's defined in terms of mating, and that doesn't happen with asexual organisms.

Furthermore, gene flow is much easier among plants and among single-celled organisms than it is among multi-cellular animals. So, the biological species concept is tailored to multi-cellular animals and might not apply more generally. Some biologists, for instance, suggest that there may be oak trees that have easier access to the genes of nearby non-oak trees than they do to those of relatively distant oak trees (which would make them members of the same species with the non-oak trees on certain gene flow conceptions).

So, the unity of a species—if these criticisms of the biological species concept are right—has to consist in more than a protected gene pool.

Like the biological species concept, *phylogenetic species concepts* define species in terms of a shared history (relevant to evolution classification system). The biological concept involves a theory about the process whereby species are created and sustained (namely, reproductive isolation). Phylogenetic species concepts are agnostic about this issue; they simply appeal to patterns of common ancestry. So, they make room for the idea that mechanisms other than reproductive isolation can lead to speciation events. So, they are silent, where the biological species concept takes a stand.

But for that reason, they provide just a *grouping criterion*; they group organisms together in ways that matter to evolution, but they don't provide *ranking criteria*. It doesn't tell us which groups, which lineages, are genera (the plural of genus), which are species, which are subspecies, which are families, et cetera. And it doesn't give us a principled way of applying these taxonomic categories. Without some story about what constitutes a speciation event, this is likely to reduce to the biological species concept. We'll get back to the notion of classification and what it aims at after examining one more species concept.

The *ecological species concept* identifies a species as a group of organisms sharing a particular ecological niche. This has caught on a little bit these days because it looks like an attractive way to handle asexually reproducing species. The idea is that asexual organisms don't compete with one another for mating opportunities (they don't mate), but they do for roles within an ecosystem. So, if something like this can be made to work, we have an analog of reproductive isolation for asexually reproducing creatures, and it could play a kind of supplementary role with respect to some of the other species concepts, for sexually reproducing creatures.

Ecology, though, is kind of a new kid on the biological block, and this approach gives a lot of hostages to fortune. It raises questions like how well do we understand the notion of an ecological niche. Does that seem like a stable, enduring, explanatory biological concept? And there's also the more direct problem of sometimes it seems that members of different species sometimes compete for the same ecological niche. So, if we just have a pure ecological species concept, we make those organisms members of the same species.

So, the fundamental question here is how important is it to unify these different conceptions of species? Monists think that there is a single correct (or at least a single best) species concept. Their hope or their assumption is that it's important to appeal to a single conception that plays a distinctive role in understanding evolutionary processes on the one hand, and in classifying organisms on the other. We need one notion to play that dual role in order to do important biological work. So, giving up on that ambition would be to abandon something central to biological thought.

The cost you pay if you adopt monism is that you run the risk of excluding genuinely explanatory species concepts. Sometimes interbreeding seems to be what's important. Great advances in treating malaria were made when biologists discovered that there were two apparently indistinguishable species of mosquito in the same geographical area, but the two species were reproductively isolated from one another. That helped them understand why malaria was common in some mosquito-infested regions and uncommon in others (the mosquitoes that looked alike were members of a different species; the way to tell that was by applying the interbreeding criteria).

But with diseases like AIDS, the valuable work is done by identifying morphology, by looking at the critter involved, by examining intrinsic rather than relational properties (like with whom something breeds). Do we want to commit ourselves to saying that there's a right way to identify organisms, or do we want to say it depends on what you're trying to do? For instance, if you're trying to prevent disease, there might be different important kinds of classifications.

Monists might try to capture everything legitimate about the other species concepts (notice this would involve a kind of ambitious reduction like those we discussed back in Lecture Twenty-Three). To what might they reduce? Something like the notion of a lineage, but it's hard to see how this is going to do justice to the range of organisms (for instance, both to sexually and to asexually reproducing ones) and also the variety of uses to which the notion of a species is put. If you try to reduce all of these to a single notion centering on lineage, will we really be able to distinguish species from subspecies on the one hand, and from higher taxa, like genera, on the other?

So, pluralists avoid this problem by saying, let 1,000 flowers bloom; there's no problem having multiple conceptions of species that answer to different biological interests. The biomedical example we just talked about is supposed to help motivate this kind of approach.

Pluralism comes in a number of degrees. Some versions are pretty limited; they require that a species be some kind of spatiotemporally continuous lineage, but hold that there are a number of equally appropriate ways of carving up the lineage, depending on what you're trying to do.

Other pluralistic approaches are more permissive than that; they allow almost any grouping that does explanatory work in biology to count as a species. That's starting to get pretty permissive. It might allow other higher taxa, for instance, to count as species.

You're relying here also on a notion of doing explanatory work that's bound to be somewhat controversial because not everybody's going to think the same things in biology need to be explained.

But there are even more permissive criteria of species, allowing non-biological notions of species. So, the kinds of the Creation scientists might count as species for some largely religious purposes. And the purposes a cook might have for distinguishing onions and garlic from one another and, more importantly, from lilies, are just as legitimate—on a very pluralistic approach—as the biological purposes that lead us to lump onions, and garlic, and lilies into a biological family.

The more pluralistic one's conception, the more it seems that one owes an account of what makes different species concepts different *species* concepts. The more different work is being done by them, the more the question presents itself: Why call all of these species *concepts*?

Furthermore, if one rejects monism, but also does not think that one can explain what makes a bunch of different lineages or groups commonly species, then it looks like the best strategy to adopt is to be a skeptic about the notion of a species, as an increasing number of philosophers of biology are. They don't deny the existence of groups of organisms, of course. What they deny is that there's anything in the world that is reasonably close to the main uses to which the notion of "species" has been put, and which can distinguish species from other biological classifications. So, there's all the biological stuff, but the notion of a species does not carve it up in a way that we should care about.

Similar issues arise much more starkly with respect to the higher biological taxa. Pluralists will stress the legitimacy of different purposes we might have in mind when classifying organisms, while monists will point out that conceptual problems are going to get in the way of many of these classification schemes. Let's see how this works out.

A phenetic classification system is going to be similar to the phenetic species concept, but applied here mainly for classification purposes. It tries to convey information about similarity among organisms in a maximally efficient way. So, the idea would be that species are maximally similar organisms; genera are maximally similar species; families are maximally similar genera, et cetera.

A related but distinct classificatory approach would be to classify organisms in terms of evolutionary disparity. This is simultaneously a historical and a morphological, or bodily, classification system. So, on this approach, a lizard species sufficiently different from the next closest species might get put in its own genus because we're trying to measure the extent to which it's disparate from other lizards.

Finally, we might classify organisms in a way designed to reveal how evolutionary history has gone. We would classify organisms in terms of their ancestry. This is called the *cladistic approach* to classification, and it's the most common approach.

Pluralists will be tempted to allow all three kinds of classification—it depends on what you want to do, what you want to convey to people using your classifying system. The monist might worry about whether these deserve to count as legitimate classifying systems. So, the phenetic system is going to run into problems about the unclarity of the similarity relation. Isn't there something peculiar about allowing that oysters and pigs could get classified together as non-kosher animals? That's not the kind of classification biologists should care about. So, the worry is that there's no principled way to measure similarity and, furthermore, that creatures that look similar to us may not be so similar (and real similarity here would be measured in terms of "the best" species concept).

Similarly, one might worry about how well evolutionary divergence can get characterized. Back in the day, somebody classifying organisms placed human beings on the level with all other mammalian species—because we're just that special. We want more objectivity imposed on our classifying schemes than this sort of notion of how disparate one set of organisms is from another imposes.

The cladistic approach, common though it is, might focus too much on descent and might not make enough room for groups that are not united by a common ancestry, but play a genuine role in explaining how evolution has gone. So, the non-marine mammals do not form a real group by cladistic standards (they don't share a common ancestral species), but evolutionary history might group them together for good reasons. Some events that matter to evolution might have happened to all, and only, non-marine mammals.

So, as we noted, cladistic classification groups together organisms that share an ancestral species. So, the notion of a reptile is not a real cladistic group because there is no species that is ancestral to all reptiles that is not also ancestral to birds.

But this notion of what it takes to have a real group does not vindicate any of the higher-level taxa. All groups that share a common ancestral species are legitimate, according to cladistic classifiers, and all groups that don't are not legitimate. A genus might name a real group, but there's nothing in reality that makes it a genus rather than a family, kingdom, whatever.

As all this talk about what it takes to have a real grouping suggests, we're touching on the same issues we talked about when we raised the distinction between "hard" and "soft" realism in Lecture Twenty-Six. Giving this some sort of biological content helps us see what's at stake, what's involved in trying to do real scientific work with the notion of classification.

The one point I'd like to make about this is it's not just a matter of imposing one's intellectual personality on the world. Monists are not fascists who want to insist that everybody do things their way. They're people who are impressed by the problems with various classifying schemes, and think we need to hold out for something better, while pluralists are people who are impressed with the legitimate explanatory use to which different classifying schemes can be put. Each of them is on to something important and internal to science. It's not a matter so much of temperament as argument.

Next time, we will turn to the philosophy of psychology.

Lecture Thirty-Five
The Elimination of Persons?

Scope:

In most cases of reduction, the entity or theory that gets reduced is still presumed to exist; we don't get rid of water by reducing it to H₂O. But in some cases, an *eliminative reduction* seems to be in order. Our best theory of demonic possession says that it never happened; every case of demonic possession is really a case of something else. A number of philosophers have adopted this attitude toward *folk psychology*, the commonsense explanation of behavior in terms of beliefs, desires, and such. Could arguments from neuroscience and philosophy of science really show that there are no such things as beliefs, desires, and persons?

Outline

I. Arguably, *folk psychology*, our commonsense approach to psychological phenomena, amounts to an ambitious explanatory theory of human behavior. It has a systematic structure deployed for purposes of prediction and explanation.

   A. The theory has an ontology: It posits unobservable states, such as beliefs, desires, emotions, and so on.
      1. The theory models thoughts on publicly observable (written or uttered) sentences. Beliefs are similarly structured.
      2. Sensations are not taken to be linguistically structured, as thoughts are. They are posited internal states modeled on external objects.

   B. Folk psychology has laws.
      1. Some of the laws are fairly closely tied to observation, for example, "People who are angry are easily irritated."
      2. Some laws are relatively distant from interpretation via observables, for example, "People will generally choose the means they believe most effective in realizing their ends."

   C. The theory can be understood using any number of tools we've developed in this course.

1. It can be structured as an axiomatic system given an interpretation through an observational vocabulary, à la the received view.
   2. Some Kuhnian exemplary applications could be added, along with some discussion of how we learn to predict, explain, and solve puzzles using this theory.
   3. We sometimes use models, as in the semantic conception of theories, to understand folk psychology.

II. Though we seem to do pretty well predicting and explaining each other's behavior using folk psychology, many philosophers think that folk psychology competes with and compares unfavorably to various kinds of scientific psychology.

   A. The laws that figure in folk psychological explanations can be saved from the "death of a thousand counterexamples" only by being protected by lots of "all-other-things-equal" clauses or by being formulated in terms of tendencies people have. They are hardly bold Popperian conjectures, and we tend to explain away their failure by appealing to ad hoc hypotheses.
   B. Furthermore, folk psychology has not made much progress in the last few thousand years. Competing research programs look more progressive.
   C. Folk psychology does not explain many phenomena that appear to fall within its domain: creativity, many kinds of learning, most kinds of mental illness, and so on. Explanations in terms of belief, desire, sensation, and such tend not to work in these contexts.
   D. Many folk psychological explanations face fairly direct empirical challenges. Split-brain experiments, for instance, suggest that we are good at convincing ourselves that we are acting for cogent reasons even when it's quite clear that we're not doing so.
   E. The entities and laws posited by folk psychology do not cohere well with those of neuroscience and other parts of scientific psychology.
   F. The categories of folk psychology bring all sorts of problems in their wake, such as how we can believe in falsehoods, see or want non-existent things, and so on. Beliefs and the things

believed are weird entities that tend to keep philosophers in business.

III. If these considerations are on the right track (admittedly a big "if"), folk psychology would seem to be candidate for an *eliminative reduction*.

    A. In our earlier discussion of reduction, we found some reasons for respecting the explanatory power of theories that don't seem to reduce to more basic theories. But that's only true if the theories do a good job in their own domain. If not, they should be replaced by a better theory, in the same way that witchcraft is replaced by a theory that posits the existence of sexism, religion run amok, and some peculiar rules of legal procedure.

    B. The failure of fit with neuroscience figures prominently in arguments for the replacement of folk psychology. Many neuroscientists think that a *connectionist* model of the mind fits the scientific data much better than does the folk psychological theory.

        1. On such a model, the brain does not "think" in states or episodes structured like sentences. Variations in patterns of stimulation across large numbers of neurons produce representations, just as variations in brightness levels produce an image on a television screen.

        2. Most information processing is, thus, subconceptual.

        3. Similarly, learning is less a matter of accumulating data stored as sentences than it is a matter of arranging stimulation patterns in the brain.

        4. The picture that emerges is not one according to which folk psychology is shallow and neuroscience deeper. The worry is that folk psychology presents a seriously misleading picture of what is really going on in our brains.

IV. An eliminative reduction of folk psychology seems to have the consequence that none of us has beliefs, desires, and such and, hence, that none of us is a person, because we think of personality through the concepts of folk psychology.

    A. Is this really possible?

1. If folk psychology is anything resembling a scientific theory, we should be open to the possibility that it is largely false.
2. It can seem problematic to deny the truth of folk psychology, because if it's false, you can't *believe* that it's false, because on that hypothesis, there are no beliefs. But this problem can be surmounted rather readily.
- **B.** The best way out of the problem is probably to defend folk psychology as a decent semi-scientific theory.
    1. The eliminativists sometimes stick folk psychology with problems that it needn't face. It is not clear that it is folk psychology's job to explain such phenomena as mental illness. It generally is used as a theory of normal, intelligent behavior.
    2. And some of folk psychology's limitations are shared by its competitors. Folk psychology doesn't have a good handle on creativity, but neither does neuroscience so far as I've been able to tell.
    3. Further, the concepts and ontology of folk psychology do seem to get used in serious and successful scientific psychology. Examples include rational choice theory, memory, and some parts of learning theory. In addition, folk psychology arguably plays a crucial role in some of the explanatory and predictive successes of history, economics, anthropology, and sociology.
    4. Perhaps most importantly, the worries about a failure of fit between folk psychology and neuroscience might be premature. We have a lot to learn in the whole domain of psychology.
- **C.** Folk psychology could be defended from elimination by denying that it is a competitor with scientific psychology. One could treat folk psychology in a strongly instrumentalist manner, for instance. We talk as if a thermostat has beliefs about the temperature in the house, but we aren't being metaphysically serious about it. Might we be talking about each other that way?
- **V.** Surprisingly but importantly, issues of what theories are and what they are for, how they are confirmed, whether they can

explain, how they fit with other theories, and so on are implicitly involved in our very sense of ourselves.

Essential Reading:

Churchland, "Eliminative Materialism and the Propositional Attitudes," in Boyd, Gasper, and Trout, *The Philosophy of Science*, pp. 615–630.

Supplementary Reading:

Greenwood, *The Future of Folk Psychology: Intentionality and Cognitive Science*.

Questions to Consider:

1. Sometimes quite a lot is at stake in a single contested term. Do you think that folk psychology should be understood as a psychological theory? Why or why not?
2. Imagine for a moment that you could become convinced that folk psychology is an irredeemably bad scientific theory. Would you give it up? What would that involve and what would be lost? To what extent is science answerable to our commonsense understanding of things and to what extent is our commonsense understanding of things answerable to science?

Lecture Thirty-Five—Transcript
The Elimination of Persons?

After our brief excursions through the philosophy of physics and the philosophy of biology, we arrive at the other main subfield within philosophy of science: the philosophy of psychology. We've actually already spent the better part of a lecture here when we looked at the possible reduction of computational philosophy (of such properties as being able to perform arithmetic inferences) to the level of the physical (to things like neural events).

Some similar issues about reduction and the mental are going to arise in this lecture, as we try to assess the scientific credentials of our commonsense psychological categories. Here, the philosophy of science touches very directly on issues central to our sense of ourselves.

We first have to raise the issue of whether we're entitled to treat commonsense psychology as anything like a scientific theory. It's certainly the case that not every theory worth treating scientifically needs to be part of a scientific discipline; it certainly doesn't need to be written down. We've seen that the positivists' rational reconstructions of scientific theories were not the sorts of things that scientists themselves used.

So, it's one thing to use a theory, and it's another to state it. So, it's quite possible that commonsense psychology is an unstated—but scientific—theory on which all of us rely. But we should address this question directly.

What makes something worth treating as a scientific theory? To a first approximation, we might think that something that has a kind of systematic structure deployed for purposes of prediction and explanation at least looks like it's a competitor with scientific theories. It looks like it's in the business of doing the kind of stuff that science does. And if it's reasonably good at its job, it's worth taking that theory, or proto-theory, with some scientific seriousness. So, we needn't commit ourselves at the outset to the claim that commonsense psychology embodies a scientific theory, but there is a sort of prima facie, an "at first blush," case to be made for the idea that it at least approximates the function of a scientific theory.

We saw Quine say that the Greek gods and atoms are in the same business as posits; they're both in the business of helping us make sense of experience, helping us predict in an economical way what's going to happen. One needn't go as far as Quine in thinking that Zeus has about the same status as electrons in order to think that a practice that invokes theoretical terms, and maybe even unobservable entities, for such purposes as prediction and explanation should count—at least initially—as a scientific theory.

So, arguably, what we'll call *folk psychology*—our commonsense approach to psychological phenomena—embodies or amounts to a really quite ambitious explanatory theory of human behavior. We will sketch folk psychology momentarily, but first we should face the question of what it might believably be if not a theory, and we'll come back to that at the end of the lecture as well.

There was a prominent literary critic five or seven years ago who had been a Freudian, then read a bit of Karl Popper's work, and then spent the next few years going around pronouncing this or that enterprise unscientific (because it seems not to have occurred to him that by strict Popperian standards, physics is unscientific as well). So, his was a game that got old pretty quickly, but he did have a good line about Freudian literary critics, which I'll paraphrase (hell hath no fury like a fallen Freudian). He said of them: "To say that you use Freud, but do so as something other than scientific psychology is like saying that you drink Sterno, but do so not as a heating element."

The application here is if folk psychology isn't to be taken seriously as a theory, what is it? It looks like a practice that is sort of ungrounded, if it's not to be taken as a serious predictive and explanatory theory. In a loose sense, science is just about trying to predict and understand the world (this is science characterized quite informally). But we haven't solved the problem of demarcation; we're not resting anything on a fancy notion of science. In an informal sense, folk psychology—at least initially—passes the test for warranting scientific consideration.

So, let's see what it would look like if we were to treat it as a more formal scientific theory.

First, the theory appears to have an *ontology*; it posits stuff. Folk psychology posits unobservable states like beliefs, desires, pains, emotions, memories, and sensations. It does this to explain observable stuff: behavior (we posit pains to explain why somebody is making a certain face or screaming, to take a very simple case).

Notice that at least initially we take these unobservables at face value. Something like scientific realism seems to pass as untutored commonsense for most of us. We're not—although we'll revisit this issue—initially inclined to think that we're merely positing beliefs and desires to make sense of observable reality. We think there *are* beliefs and desires. So, perhaps, it takes some philosophical or scientific training to beat an empiricist's temperament into one. We start off as realists. If our theory says that there are beliefs and desires, we seem to think there are beliefs and desires, even if we don't know exactly what they're supposed to be. We at least talk as if we take the existence of beliefs and desires seriously.

Well, what are they supposed to be? Keep in mind that I'm not speaking with any special expertise here. This is supposed to be your psychology as well as mine; it's commonsense property. But it's at least plausible to suggest that, as standardly understood, the theory posits thoughts as being modeled on publicly observable sentences (written sentences or spoken sentences). *Beliefs* are enduring versions of thoughts. To say that somebody is thinking something is to say something like he or she is thinking that it's humid outside today. That's a sentence-sized thing, and if it sticks around, it's a belief rather than merely an occurrent thought. So, thoughts and beliefs are sentence-sized and sentence-structured. They are, as it were, modeled on sentences—written or spoken.

Sensations are not linguistically structured like thoughts; they are posited for different explanatory purposes within commonsense psychology. They're inner states modeled on external objects. Roughly, they're the ideas in that technical sense we saw in Locke, Berkeley, and Hume. When I see my dog playing in my backyard, what I have is a little mental copy of that scene. That little visual copy is a sensation and, if abstraction is a legitimate operation, then so are its components; I can have a sensation of my dog's color, for instance.

So, that's a partial look at the ontology of folk psychology. Like any good or not-so-good theory, it has laws or proto-laws as well. The theory can be represented more or less as the "received view" of scientific theories would have it—that is, as a set of sentences including laws and issuing in explanations and testable predictions.

So, some of these laws are relatively "low-level." That means they're reasonably closely tied to observation. An example of such a law would be "People who are angry are easily irritated." An analog of this in a more paradigmatically scientific theory would be a relatively low-level law like "All copper conducts electricity." Notice that this low-level psychological law can be treated also as a partial interpretation of the theoretical term "anger" in observation terms—if we grant, at least for the sake of argument, that irritation is an observable state. If not, we'd have to find an observable state in which we're interpreting a notion like anger. And other low-level laws involving anger like "People who are angry tend to raise their voices" also help interpret the theoretical terms of the theory observationally.

In addition to these low-level laws—laws that are reasonably closely tied to observation—there will be higher-level laws, which are more fundamentally tied to explanation than to observation. Higher-level laws are distant from interpretation via observables. So, for instance, a law like "People will generally choose the means they believe most effective to realizing their ends" is not very closely tied to observation. That's a theoretical statement. It might be an axiom of the theory; it might not derive from anything more fundamental—it might be an unexplained explainer. Such "high-level" laws will help explain other lower-level laws (like "All other things equal, people will take the job that pays the most money").

We use these laws tacitly (we don't generally state them out loud or to each other), combined with suitable initial conditions, to explain a vast and rich range of human behavior. The examples I've been giving are comically simple. Philosophers tend to use kind of ridiculous examples for the purposes of clarity, but I don't mean to condescend to folk psychology as an explanatory theory.

In the hands of a good observer of human behavior and a good theorist of human nature—like some biographers, for instance—one gets really quite interesting explanations of apparently diverse patterns of behavior. A person's life can get unified around a few interesting theoretical posits. How testable, how correct, these explanations are is a further question, but this is not a toy theory, though I've been talking as if it was because I wanted to get the basic ideas of the theory out there.

And, of course, we use this theory to predict one another's behavior and to explain one another's behavior. The crew here at The Teaching Company by now probably has a rich store of explanatory and predictive knowledge that it uses quite effectively to deal with homo academicus—they know how to handle professors around here. They know what kinds of caffeine and throat lozenges to keep around, and how to deal with the academic ego. And, in an everyday sense, we use this theory in dealing with our partners, our children, our parents, our friends, our enemies. In order to know what to expect from other people, we explain their behavior in terms of what they're seeing, sensations, what they're thinking, what their intensions on the basis of what they're seeing and thinking plausibly are.

Despite the fact that we all tend to be realists about the ontology of folk psychology (we talk as if beliefs and desires really exist), the positivists would much have disapproved of that—they tended to prefer psychological behaviorism, which reduces, in the philosophical sense of "reduce," unobservable mental states (like beliefs and desires) to observable behavioral ones. Most of us don't go in intuitively for a positivistic reduction; we think beliefs and desires have a kind of reality.

Nevertheless, I've presented the theory so far in the classic positivist fashion, with the theoretical terms getting partial interpretations through observables—both to kind of remind us of this central aspect of the course, and to show how the theory might get a kind of familiar structure and how meaning might flow into the theory from observation.

But this is not the only way that we could render folk psychology as a scientific theory. Arguably, we might want to introduce some Kuhnian elements. So, Kuhn thinks that a paradigm (which is loosely identical to a theory) needs exemplary applications of the theory; it needs a set of values that animates the explanatory and predictive projects of the theory and a kind of vocabulary that sets the problems and allows normal science to happen. Notice that normal science here, informally construed, is figuring out what each other are going to do.

We don't tend to see much of that within normal science, but we do see it in sort-of specialized versions of normal science. Freud, for instance, uses the categories of folk psychology and adds a sort-of specific ontology of his own—involving the id, the ego, and the superego—within a kind of folk psychological framework. Freud offers paradigms, in the Kuhnian sense—cases studies that one is supposed to read and learn how to apply the theory. The notion of a "Freudian slip" is a kind of paradigm of specialized folk psychological theory.

The way we learn folk psychology probably explains the fact that there aren't exemplary applications and things like that. We don't learn this in an academic context; we learn it as children. So, we don't need paradigms to be taught to us, to guide the application of the theory later in life. It's an interesting question to what extent we could muster paradigms, exemplary applications, normal science solutions, to guide this sort of explanation. We don't seem to need them, but if we had to come up with them, it's an interesting question what they'd look like.

We could also deploy some of the resources of, say, the semantic conception of theories, and could interpret folk psychology through models rather than as a set of sentences ordered in an axiom system. Arguably, I understand other people by modeling them on myself. This is called a *mental simulation approach to the mind*.

It could provide an interesting test case for the semantic conception of theories, but it's going to raise some complicating issues that we don't want to go into here. Many philosophers *contrast* understanding other people through simulation and understanding them through theorizing about them. So, on their view, simulation doesn't provide a theory; it provides a sort-of more direct kind of access to what's going on with them. So, it's not clear that simulation is a model in the sense that matters for the semantic conception of theories.

Going back to kind-of baby examples, we do, in fact, use models to interpret folk psychology: Think of the kind-of classic case from Saturday morning cartoons, where a character has a devilish version of himself appear on one shoulder and an angelic version of himself appear on another shoulder. This is a kind of quasi-Freudian interpretation of the id and the superego, but it's a model—it's a toy model, but it's a model that gestures at a kind-of explanatory picture of human behavior.

If we depart from the received view of scientific theories, then our explanations might be offered in terms of these models or pictures rather than in terms of laws, but that needn't make them any less explanatory. It would be a problem for a covering-law approach to explanation, but not a problem for explanation as such, and the covering-law view can be modified to try to take account of these complications.

We'd like to think we do pretty well predicting and explaining each other's behavior using folk psychology. But many philosophers think that folk psychology competes with—and compares unfavorably to—various kinds of scientific psychology. Let's just run through some of the major worries about folk psychology.

The laws that figure in folk psychological explanation face the "death of a thousand counterexamples." Many times when people are angry, they get quiet—not loud—for instance. So, these laws—in order not to be falsified—get hedged by lots of ceteris paribus clauses, by lots of "all other things equal" clauses, or by being formulated in terms of tendencies people have rather than behavior they will exhibit.

So, the laws of folk psychology are hardly bold, Popperian conjectures. They look, in some ways, more like psychology: predictions that are vague and that are protected from falsification by sometimes some ad hoc maneuvers. So, it's not clear, for instance, that we treat a hypothesis like "So-and-so is just being shy" as an independently testable explanatory hypothesis about behavior, rather than just as a way of protecting an utterance or prediction we just made from having been falsified.

So, the worry is the theory looks more successful to us than it is, just as astrology looks more successful to its practitioners than we think it is because we unselfconsciously protect its claims from apparent falsification. Furthermore (and developing Popper's view in the direction that Lakatos developed), folk psychology doesn't seem to have made much progress in the last few thousand years. Critics claim that competing research programs within scientific psychology look enormously more progressive than folk psychology has.

Folk psychology doesn't appear to explain some of the central phenomena within its domain. It doesn't have much to say about how human creativity happens, how language learning happens, how certain kinds of mental illness come about. Explanations in terms of belief, desire, and sensation seem not to work very well in these contexts.

Furthermore, many folk psychological explanations face rather direct empirical challenges. Experiments with split-brain patients, for example (these are patients in which one hemisphere is unaware of what's going on, as it were, in the other hemisphere), seem to suggest that a lot of what look like folk psychological explanations or predictions are, in fact, rationalizations. One hemisphere will find itself observing what the body is doing (and this is explained by some suggestion or perception going on in the other hemisphere), but the hemisphere in question will say, "Oh, no, I wanted that drink of water, I'm thirsty," even though we know that the hemisphere that's saying this wasn't aware that the water was there. We rationalize, we make sense of our behavior, but we're not explaining how our behavior is caused, though we think that we are.

We have a kind of psychological need, on this picture, to see our behavior as caused by reasons, but it's not clear that this is generally true. It's often a rationalization rather than a prediction.

The deepest problem, many think, is that the entities and laws posited by folk psychology just don't fit very well with those of neuroscience and other parts of scientific psychology. We saw this very problem when we looked at attempts to reduce the mind's computational functions to brain states (and we'll see a different worry about reduction in just a few minutes).

Finally, though, categories of folk psychology seem to bring all kinds of problems in their wake. Belief and desire raise a whole bunch of philosophical problems about, for instance, how we can want things that don't exist, how we get to think about things like that. So, it's a psychological theory that seems to keep philosophers in business, and that's almost never thought to be a virtue in a scientific theory. Some people think that the very language of folk psychology raises explanatory problems that the language itself makes hard to solve.

So, if the above considerations are on the right track (admittedly, that's a big "if"), folk psychology looks like a candidate for what we called an *eliminative reduction*, when we looked at reduction back in Lectures Twenty-Three and Twenty-Four. Back then, we assumed that the theories to be reduced did a pretty good job of explaining and predicting within their own domain, and we found reasons for respecting the explanatory power of theories that don't seem to reduce to more basic theories. It seems like we sometimes lose explanatory power when we go down a level to something more fundamental.

The worry here is such explanatory power as some theories possess does not seem to make it tempting to salvage the theory's laws or ontology. The theory that witchcraft happens is one that we don't think explains enough to allow us to reduce it to sexism, religion, and peculiar rules of legal procedure. The idea is it's much clearer to say witchcraft gets eliminated, and the explanation proceeds via these really quite different categories. Phlogiston does not reduce to oxygen, but gets replaced by it; witchcraft gets replaced by, say, psychological or historical explanations.

So, some philosophers think that folk psychology is in the same boat, though I hope they would admit that it's not as bad a theory as the witchcraft theory is. The failure of fit with neuroscience, as I indicated before, looms especially large in these arguments.

Many neuroscientists think that a *connectionist* model (I'll deal with this only in the broadest of terms) fits the mind better and the scientific data much better than does the folk psychological theory. On this approach, the brain doesn't "think" in states or episodes that are at all structured like sentences. Thinking happens when variations in patterns of stimulation across large numbers of neurons produce representations, kind of like variations in brightness levels produce an image on a TV screen. There's nothing linguistic underlying that. The processing is subconceptual—it's a pattern of distributed excitations rather than a thought that something is the case. There's nothing linguistic-ish or linguistic-sized going on in the brain on this picture.

Similarly, learning isn't a matter of accumulating data stored as sentences so much as it is a matter of arranging stimulation patterns in the brain. That's how categories get embodied: by a difference in electrical potential. So, such an approach is pretty well documented when we want to explain, for instance, how babies recognize faces. They don't have language yet, but they're able to recognize faces because the explanation here is connectionist—not semantic. So, things like pattern recognition—as embodied in differential stimulation patterns—presents a pretty different picture of the mind than folk psychology does.

So, some philosophers conclude that there's nothing in the brain answering to the beliefs and desires that folk psychology posits. They further claim that folk psychology is not a good enough theory to stand on its own, to overcome the failure of fit it exhibits with neuroscience.

So, this case looks rather different than the case discussed in Lecture Twenty-Four, where the reduction of computational psychology to the physical was disjunctive and messy, but not completely unpromising. Here, the idea is the reduction is nonexistent. Reducing folk psychology to neuroscience would be like reducing witchcraft to psychology, and sociology, and history, and religious explanations. The idea is not just that folk psychology is shallow and that neuroscience is deeper, but that folk psychology presents a seriously misleading enough picture of what's really going on that we should get rid of it.

An eliminative reduction of folk psychology would have the consequence that none of us really has beliefs, none of us has desires, and none of us is a person (where that means the kind of thing that thinks—and thinks by using beliefs, desires, that sort of stuff). Could that really be the case? Could neuroscience show us that there are no beliefs, and that you and I are not persons?

If folk psychology is anything resembling a scientific theory, we need to be open to the possibility that it is—at least in large part—false. The eliminativists are suggesting that folk psychology is like astrology or like other fundamentally misguided theories. It's hard to contemplate giving it up if it's integrated deeply into your web of belief, but if you step outside of it and look at it from a neuroscientific perspective, the theory seems fatally flawed.

If it's used for prediction and explanation, folk psychology looks to be making claims about the world, and so those claims had better be substantive, predictive claims, and if they are, then they're falsifiable—or at least so it seems. At the end of the lecture, we'll revisit this assumption.

There's a kind of technical problem here. Arguably, it's unintelligible to deny the truth of folk psychology. Why? You can't believe that the theory is false because if folk psychology is false, there's no such thing as a belief. But there's some reason to think that we can use a theory against itself. If we found what we'd be tempted to describe as a pre-Cambrian rabbit, that might bring about enough changes in our geological and biological theories that we would end up jettisoning the concept of the Cambrian period. But that doesn't make this string of inferences incoherent; it means some evidence described in the theory's terms ends up overthrowing the theory.

So, the most straightforward way out of the problem would be to defend folk psychology as a decent scientific—or semi-scientific—theory, and it's not at all clear that that's an unpromising task.

So, for one thing, some of the failures of which folk psychology is accused might not be part of its job. It's not clear that folk psychology is in the business of trying to explain mental illness, that it's supposed to be explaining dreaming or the function of sleep, or how language is acquired.

Why? Because folk psychology is often put forward as a theory of normal, intelligent behavior. As a theory of normal, intelligent behavior, it is not committed to claims about perception; it's not committed to claims about mental illness. Those might be susceptible of a kind of neuroscientific explanation. When we're trying to explain each other's everyday behavior, we needn't have a built-in theory of mental illness at work.

Eliminativists think that's a misguidedly narrow approach to psychological explanation. They think it's like trying to develop a chemistry about a few elements that hang out together in the middle of the periodic table.

But non-eliminativists might say that it's more like trying to develop a theory of light, which is only one kind of electromagnetic radiation, among others, but it's one that can be largely—if not entirely—understood on its own terms, and it's something well worth having its own theory of. So, if normal, intelligent behavior is different enough from the psychology of sleep or of mental illness, then it's worth having its own theory.

Folk psychology could be broken up into distinct theories (we could have a theory of belief, a theory of pain, things like that). If those phenomena are different enough to require a different ontology or different laws, then folk psychology could divide itself up. We want to make sure, though, that it doesn't contract merely to protect itself from falsification. We want the theory to be reasonably bold and falsifiable; otherwise, it can't be a good scientific theory.

We have to make sure we don't judge folk psychology by unduly harsh standards. Folk psychology does not have even the beginnings of a good explanation of human creativity—but neither does neuroscience, really. So, to criticize folk psychology for that particular failing is mostly a matter of saying psychology is a very young science.

And the concepts and ontology of folk psychology do, arguably, figure in some successful scientific psychology. Rational choice theory, the theory of memory, some parts of learning (not language, but some other parts of learning) do get handled—in part—by appealing to concepts, beliefs, and intentions. This could get explained away, but if these approaches are at least reasonably successful, then we could make the claim as we did about explanations at the ecological rather than the biological level, that these notions are pulling their own explanatory weight, and so needn't get reduced to some more basic level.

Furthermore, folk psychology, arguably, should get credit for part of the explanatory and predictive success that history, economics, anthropology, and sociology have attained because those sciences are just pervaded by folk psychological vocabulary and rely implicitly on those laws.

So, if folk psychology is a pretty good theory, then we might not want to lose explanatory power, even if it doesn't reduce very well to the neurological level. We saw the same thing with computational psychology. We don't want to reduce the ability to add; we don't want to insist that the ability to add get reduced to the physical level because we lose too much explanatory power.

The pessimism about the compatibility of neuroscience with folk psychology might be premature. Psychology as a scientific discipline is only about 100 years old. We've got a lot to learn still about how the brain works, and a lot to learn about folk psychological explanation. So, it's far from obvious that the failure of fit has been established. We don't see how folk psychology fits into neuroscience, but that doesn't mean we see that it doesn't.

Finally, folk psychology could, in fact, function rather differently than as a scientific theory. It could be a kind of useful fiction or, in the language of this course, it could get an instrumental interpretation. So, the philosopher of mind, Daniel Dennett, treats folk psychology instrumentally. We adopt a kind of predictive and explanatory stance with respect to some creatures and objects rather than others. We can use it with respect to a thermostat. We can treat the thermostat as trying to regulate the temperature in the room, even though we don't really think thermostats have goals. It's sometimes useful in understanding a system like that to talk as if it has beliefs and desires.

So, on Dennett's view, we take this approach with one another; we talk as if beliefs and desires are real things, but, in fact, it should get a kind of positivistic construal. Belief, and desire, and categories like that are placeholders; they're calculating devices. They're not assertions about how unobservable reality goes, but they're ways of coping with observable phenomena.

So, we see that it's not clear that folk psychology needs to be a scientific theory in order to be—in some sense—legitimate, and if it is a psychological theory, we see that it doesn't have to have its status as carving the world at its joints convincingly and conclusively established. We saw that it's not completely clear what answers in biology to the notion of a species, but species skepticism is still a minority opinion. So, even if we don't quite know the place of belief and desire in the world, if they look like they might pull their explanatory weight, we might still take them with some ontological seriousness.

So, we've seen another example about what theories are for and what we're committed to in taking their ontology seriously. The very sense we have of ourselves depends—in part—on what we take a scientific theory of creatures like us to be in the business of doing.

Next time, we will try to bring it all back home.

Lecture Thirty-Six
Philosophy and Science

Scope:

In this lecture, I will attempt to forge some new connections among our by-now-old ideas. Our overarching themes involve tensions between tempting ideas: the distinctiveness of science versus the continuity between philosophy and science, the competing modesties of empiricism and realism, respect for accurate descriptions of scientific practice versus the legitimacy of attempts to improve the practice, and the importance of developing a picture of the intellectual virtues and values of science and scientists that is neither cynical nor smug. The lecture (and the course) aspires to leave you puzzled in articulate and productive ways.

Outline

I. We began the course by wondering what is special about science. The idea that there is something distinctive about the sciences is attractive, but it sits awkwardly with attractive aspects of holism and naturalized epistemology.

   A. It is not clear to what extent we want to distinguish scientific from everyday theorizing. We do not take ourselves to be doing science in our everyday lives, but we properly aspire to embody scientific virtues in many of our everyday undertakings.

   B. Just as science had better be different from our everyday practices but not too different, science had better be different but not too different from philosophy. We've seen a number of reasons for thinking that science and philosophy should be thought of as continuous, but we mustn't lose sight of the distinctive manner in which science manages to put questions to nature.

   C. The search for a demarcation criterion, however, does not look promising. Science probably cannot be done without some kind of metaphysical picture or conception lurking in the background.

   D. The inescapability of metaphysics emerges most clearly in notions of categories, kinds, properties, and so on. What

counts as two things being similar? Which terms can figure in laws? It's often hard to see what would count as an adequate defense of our category schemes.
- E. How can these tensions between the distinctiveness and continuity of science be resolved or, at least, softened?
  1. We should recognize that our metaphysical views are not very readily tested by the world; thus, we should be modest, flexible, and self-conscious about them.
  2. We should think of science as differing from other pursuits in a number of medium-sized ways, rather than in one big way or in no way at all. Science involves a distinctive combination of observation, education, social structure, and other elements.
- F. Much of the best philosophy of science these days reflects the continuity between philosophy and science because it is both informed and driven by empirical concerns. Quine's holism and naturalism help us to see that we are sometimes working on the same parts of the web from different angles. But this holism can make the task of distributing criticism across the web of belief challenging.

II. Empiricism, both about meaning and about evidence, is an attractive idea, but it is difficult to keep it in check, and it sits poorly with scientific realism, which is an attractive idea.
- A. Empiricism about meaning is particularly unfashionable with philosophers these days—and for good reason. It hamstrings our ability to talk about almost anything. But the lesson of special relativity still looms. The greater we extend our semantic search, the more risk we run of exceeding our epistemic grasp. Empiricism helps us avoid muddling our meanings.
- B. Scientific realism seems compelling, but a realist needs to stay in touch with his or her inner empiricist. Inference to the best explanation is fragile under favorable conditions, and if we are going to be realists, we should squarely face the limitations of our evidential situation.
- C. Realists need to remind themselves of the epistemic risks they run and should try to be as clear as possible about the intellectual benefits of the risks. For their part, empiricists

need to remind themselves of the intellectual (for example, explanatory) resources of which their empiricism deprives them and should strive to be clear about the benefits of their relative asceticism.
1. If you stick closely to what is given in experience (and do not assume that too much is given), you will avoid certain mistakes. Popper's skepticism about induction falls into this camp, as does resistance to Bayesian subjective probabilities. But increased security comes at the cost of diminished resources and an increased vulnerability to skepticism.
2. On the other hand, such things as explanatory ambitions, models, and analogies and a willingness to take subjective probability assignments seriously allow one to set out to maximize the range and depth of one's beliefs. But here the risks are muddleheadedness and mistakes.

III. Kuhnian fidelity to actual science is an attractive idea, but it sits awkwardly with the "is/ought" distinction.
  A. The smart money is on scientific practice over philosophical advice about scientific method. But what scientists say they're doing doesn't always reflect what they're actually doing. Scientists tend to commit philosophy when they explain what they do. And we've seen a number of respects in which sympathetic observers might think that scientific practice might be improved.
  B. Such terms as *objectivity* can be dangerous because they tend to lead people to exaggerate the virtues and/or the vices of science and scientists.
    1. People sympathetic to such views as social constructivism hear talk of objectivity and picture scientists claiming to view nature from nowhere, to step out of their skins, to carve nature at its joints, and so on. They rightly regard most of this as pretty naïve, but that stems from the too-demanding notion of objectivity being deployed.
    2. People sympathetic to realism and/or empiricism hear scoffing about objectivity and think that science is being reduced to mere rhetoric or worse. They then tend to dismiss legitimate questions about, for example, the role

of values in science. This leads to soft-headed thinking about such things as the problem of demarcation.
   C. Science deserves a distinctive kind of respect. No amount of examining it, warts and all, undermines its achievements. But we laypeople should not accord it automatic deference.
IV. Philosophy, especially philosophy of science, is hard. It compensates us only with clarity, with the ability to see that the really deep problems resist solutions. But clarity is not such cold comfort after all. As Bertrand Russell argued, it can be freeing. When things go well, philosophy can help us to see things and to say things that we wouldn't have been able to see or to say otherwise.

Essential Reading:

Godfrey-Smith, *Theory and Reality: An Introduction to the Philosophy of Science*, chapter 15.

Supplementary Reading:

Rosenberg, *Philosophy of Science: A Contemporary Introduction*, chapter 7.

Questions to Consider:

1. Which has changed more as a result of this course, your conception of science or your conception of philosophy?
2. Which of the tensions sketched in the last lecture do you think it would be desirable to resolve and which ones seem essential to the success of the scientific enterprise (and, hence, need to be left unresolved)? Can we, when doing serious intellectual work, simultaneously treat something as a puzzle or problem yet think it is better left unsolved?

Lecture Thirty-Six—Transcript
Philosophy and Science

I'd like to begin this last lecture by expressing my sincere admiration for those who have made it this far. You have received a moderately massive dose of philosophy and of science, and if this is the way you choose to spend a good chunk of your leisure time, you're my kind of person. I admire anybody who has made it this far.

I have not generally thought it's been my job in this course to give you my opinion about the material we've been discussing because I don't honestly think you have much reason to care what I think about it. But I will, in this last lecture, take the liberty of making some suggestions about themes that have emerged from the course. I mean them seriously as suggestions, but I seriously don't mean them as more than suggestions. I find this material enormously difficult, and I find it difficult to be articulate, much less decisive about these matters. So, my suggestion will be that the most interesting themes are tensions between individually attractive positions. I'm not going to tell you what I think the right answer to these questions ends up being, so much as the right way to try to be articulate about how to face the difficulties that these questions pose for us.

Intermittently, I think some tolerably clear morals emerge from this story, but, by and large, what I think we learn is how to inhabit profound intellectual tensions successfully.

We began the course by wondering what is special about science. The idea that there's something distinctive about the sciences is attractive, and remains attractive, but it sits awkwardly with attractive aspects of holism and of naturalized epistemology.

Like many of our questions, this one has actually morphed into several reasonably distinct questions. The empirical sciences are different, not just from pseudosciences and their ilk, but also from philosophy, from everyday theorizing about the world, and from other worthwhile enterprises. And those are different distinctions that might have to be unpacked rather differently one from another.

As saw in the last lecture, it's not clear to what extent we want to distinguish scientific from everyday theorizing. This is, itself, one of those tensions that resist a clear and decisive answer. Folk psychology probably does not aspire to the kind of explanatory depth we associate with scientific theories—it's a theory (if it's a theory) driven as much by a need to cope with one another as by some aspirational project for unifying or explaining what's really going on with one another.

But it's dangerous to say that folk psychology is in a different business from scientific psychology, or even from neuroscience, in part because the function of such a claim is largely to deflect serious criticism that might be aimed at the laws or ontology of folk psychology. So, maybe we don't take ourselves to be doing full-fledged science in our everyday lives, but we do need and want to hold ourselves, in our everyday lives, and not just about folk psychology, to many of the standards of science.

So, this issue arises in many of our everyday explanatory enterprises. We don't give rigorously scientific explanations most of the time—but scientists don't give rigorously scientific explanations most of time. This is the notion of a rational reconstruction that has figured so prominently in this course.

But though we don't hold ourselves to highly formal or highly ambitious standards of explanatory explicitness, nor do we want to countenance a drastically different and lower standard than the scientific one for our ordinary explanations. We might, if we're thoroughly persuaded by Bas van Fraassen's deflationary account, according to which nobody really explains anything; we just answer one another's "why" questions. That's a contender. But assuming that one takes the explanatory enterprise seriously somewhere, we don't want to draw too stark a distinction between a successful scientific explanation and a successful non-scientific explanation. Because science, broadly speaking, is supposed to represent us at our epistemic best—it's not something deeply different than what we're doing in our everyday projects of trying to understand the world.

So, science had better be different from our everyday explanatory and predictive practices, but it had better not be too different from them. Likewise, it had better be different, but not too different, from philosophy. Fraught though the process is, and as hard a time as we have had over and over characterizing the way in which observation and evidence distinctively bears on science, science does find a distinctive way of putting questions to nature and getting at least tentative answers from nature. Philosophy works differently. There's a difference that makes a difference between what philosophers do and what scientists do.

On the other hand, the search for a demarcation criterion (or a useful bundle of demarcation criteria) did not work well, and it does not look promising. So, philosophy, and science, and metaphysics don't look radically distinct from one another. There seems to be at least that much truth in philosophical naturalism.

In particular, science probably cannot be done without some kind of metaphysical picture or conception lurking in the background. Desperate attempts to avoid metaphysics—of the sort we've seen in logical positivism most prominently—run the risk of leading to bad metaphysics and to bad faith about metaphysics (by which I mean a refusal to admit that you are perpetrating metaphysics when, in fact, you are). It's very hard to come up with a way of doing science that involves no metaphysics. The positivists had views about what's really real (it is very, very closely tied to observation, for instance)—but they had a hard time admitting that, because they didn't want to make metaphysical statements.

The inescapability of metaphysics emerges most clearly in the notions of categories, kinds, possibilities, and groups. These are inescapable in scientific and in everyday explanatory practices. In any given context, what counts as two things being similar, two emeralds having the same color (is that "grue" or "green"?), two organisms belonging to the same species, does that depend on our purposes, or is that settled by the world?

What counts as being rationally and properly indifferent between somehow similar possible states of affairs? In our situation of ignorance, how are we to represent different ways the world might be? We categorize; we impose. Our language, our culture, our biological nature, our history dispose us to clump things together. We cannot do without these categories, wherever they come from, but for the very reason that we can't do science without them, it's hard to give them an independent scientific vindication.

Our commitment to these categories gives us a kind of inner realist in our intellectual personality—someone who takes these kinds and categories seriously, who uses them to explain and to understand, who is not distanced or alienated from what the theory seems to commit us to saying about the way the world is. That's a genuine and valuable part of one's intellectual or scientific personality, but it needs to be balanced by an inner anti-realist who explains these kinds as ours, as arising contingently from aspects of our nature that might be misleading. Even our biological nature, for instance, might predispose us towards misguided psychological assumptions or assumptions in physics (like, for instance, the Euclidian geometry of space).

So, even if a category seems deeply natural and all but inescapable to us, our inner anti-realist wants to raise questions about whether we should try to do without it.

There's no general answer, in my view, to the tension between realism and anti-realism. The relevant virtue is not maintaining a consistent realistic attitude or a consistent anti-realistic attitude. It's transparency and honesty. We want to be aware of the correlative dangers of dogmatizing and of deflating.

How are these tensions between the distinctiveness of science and the continuity between science and other enterprises to be—if not resolved—at least softened? First, we should realize that our metaphysical views, our categories that we bring to the table in doing science, are not very directly testable—and therefore, we should pursue modesty and flexibility with respect to these categories, with respect to our sense of what's possible, of what's a real kind, of ways in which two things are similar.

We should recognize that science differs from other pursuits not in being free of prejudices—not in the way the world imposes itself directly on the scientist's mind, but rather in a lot of medium-sized ways rather than in one or two big ways. This is how we can maintain some genuine distinctiveness for science, without claiming to solve the problem of demarcation with which we spent so much time early in this course.

Science does involve a distinctive combination of observation, of education, of the Kuhnian indoctrination sort, among others. It has a distinctive kind of social structure. What's special about science is the way that it manages to balance Popperian criticism (subjecting one's views to criticism from other people and—speaking metaphorically, but importantly—from the world) with a kind of Kuhnian confidence and assurance that allows science to build, to assume that it's gotten parts of the world figured out, and to do something using these cognitive resources. It is a delicately maintained balance, and it's an imperfectly maintained balance. Science probably should not be credited with having gotten this right remotely once and for all. It's something that needs to be rethought, and the history of science helps show us ways in which it's been importantly rethought.

Reflecting this continuity between science and other worthwhile enterprises, most of the best philosophy of science these days is both informed by and driven by empirical concerns. I hope you got a sense in the last three lectures before this one that the richer, more scientifically informed approach to the general philosophy of science topics is rewarding. The topics raise new and interesting questions and inform more abstract issues about, say, what it takes to think of a kind as real, that allows philosophers to offer things to scientists. Scientists are starting to take philosophical work a bit more seriously than they had because philosophical work is more scientifically informed than it had been.

General philosophy of science, of the sort we spent most of the course doing, does not kick in very directly in laboratory work, but I think it can—in a bunch of ways that will emerge in the course of this lecture—help scientists do their day jobs. It helps them be articulate about the advantages and dangers of different approaches to their discipline. That's different from the work that is very informed about what a species might be or how thermodynamics impacts views about the direction of time.

The holism and naturalism of somebody like Quine help us see that philosophers and scientists can sometimes be working on the same parts of the web of belief, but approaching them from different angles. So, we can recognize that philosophy and science are not identical undertakings, but that they could be part of what is genuinely, in some sense, one enterprise.

This very holism, however, points to another tension, which is the difficulty of figuring out the right way of assigning praise and blame across the web of belief. We don't want to be pure relativists about this and just say anything that's not forbidden by the laws of deductive logic is perfectly fine, but nor did we have much luck coming up with, say, an inductive logic that tells us the right way to apportion praise and blame.

At best, if we're relying just on what the world can do, the world will tell us that something has gone wrong somewhere, and it's up to us to be bold, honest, and careful in trying to figure out how to modify our web of belief in the light of experience. We have to realize when we are, perhaps, excessively attached to some pet theory, and that we should consider letting that be the part that gets falsified by what experience seems to kick up. The world won't settle these sorts of questions for us, but that does not mean—as Kuhn and some of his constructivist followers sometimes seem to suggest—that the world gets no say, that this is a free decision on our part.

It's a constrained decision, but it is, as it were, a freely constrained decision. If we do science properly, *we* let the world tell us what to do, and that's the kind of freedom we want in doing science. We don't want license; we don't want to be able to do whatever we want. We want to have some control in how we let our beliefs be informed by experience.

We now turn to another tension in science. Empiricism—both about meaning and about evidence—is an attractive idea, and it dominated the first part of our course. But it's difficult to keep empiricism in check, and empiricism sits rather poorly with scientific realism, which is another attractive idea that tended to dominate the latter part of our course.

Empiricism about meaning, the idea that what we mean by our terms cannot be allowed to outrun observational checks on meaning, is particularly unfashionable with philosophers these days—and for good reason. We've seen that it hamstrings our ability to talk about anything that's not fairly, directly presented in experience. That dooms important parts of the scientific enterprise and maybe of everyday explanatory enterprises.

But let's not lose sight of the lesson of Einstein and special relativity. Once we start appealing to models, to analogies, to other ways of letting our terms get meaning, our semantic reach starts to outrun our epistemic grasp. We lose some valuable constraints that inform, as well as constrain, what we're saying. We don't want to start thinking that we understand what absolute simultaneity might mean. So, if we stop construing the positivists' doctrines about meaning as rules that one must follow in order to do science, and start thinking of them as reminders about some important scientific values, empiricism about meaning can be rehabilitated, and is a valuable scientific doctrine.

Because we don't want to let our meanings get muddled. Constraining our meaning in observational terms is a nice way of avoiding muddle—and who knows better the dangers of conceptual muddling than philosophers? A crucial part of our valuable function is to help scientists stay clear about what they're saying. So, empiricism about meaning is, I think, more valuable than a lot of my colleagues find it.

Nevertheless, I incline towards scientific realism, which requires rejecting a strict empiricism about meaning. It also goes beyond empiricist strictures on the use of evidence. So, for instance, I'm willing to use—in many contexts—inference to the best explanation as evidence for the truth of what a theory says about unobservable reality. So, that's almost sufficient, anyway, for making me a scientific realist. But we realists need to stay in touch with our Inner Empiricist, another part of our intellectual personality, and especially our Inner Empiricist about evidence rather than about meaning—because inference to the best explanation is fragile even under favorable conditions. Even if we help ourselves to the idea that, without some argument, we can nevertheless trust inductive inference—we never answered Hume's skepticism about induction. Inference to the best explanation is inference that goes not just beyond observation—but beyond the observable. We should—especially if we're realists—remain perpetually uncomfortable with that. We should squarely face the limitations of our evidential situation.

Here, Kuhnian talk of values—rather than positivistic talk about rules or methods—is helpful. When we go boldly beyond experience, as we must if we are to pronounce something a law in the intuitive sense (if we want to say that being made of copper makes something conduct electricity, if we want to say that's not just the sort of thing that happens, but it's the sort of thing that must happen), if we want to say that unobservable reality is a certain way, even though we can't directly detect that it's that way. If we want to evaluate counterfactual conditionals, if we want to claim to know how the world would behave if things were somewhat different than they are. These are all valuable things to claim oneself able to do.

We should be clear exactly what benefit we're getting from these leaps beyond the evidence. There's something intrinsically, epistemically presumptuous about realism. Being presumptuousness is not always automatically a bad thing to do, but we should try to explain to ourselves why we think we're entitled to the presumption. Do we think, literally and straightfacedly, that the world will answer to our explanatory ambitions, that the fact that something unifies our conception of the world is evidence that it's true? If we don't think that, do we have a different story about why it's okay to perform such inferences? Do we have a somewhat anti-realist story, for instance, according to which all we're trying to do is come up with a unified theory about the world, and we're not worrying about truth in some robust correspondence sense to the world?

On the other hand, if we're going to do without laws, if we're going to do without deep explanations, if we're going to do without claiming that there's some necessity out in the world that we can detect, we have to admit it and actually do without them—not just sanitize them into a pretend empiricist version of them, but then appeal to them in our everyday, unreflective lives. So, if we're going to go beyond the evidence, we have to admit the epistemic costs. If we're going to constrain ourselves tightly within the evidence, we have to admit the costs to our explanatory ambitions.

So, a great deal of the philosophy of science is about a kind of general ethics of epistemic resource management. That's a sort-of dull, ponderous phrase—it makes it sound like a business course. But my point, I think, is an important one: Science and philosophy can work wonders, but neither can work magic. What we're able to get out of these disciplines will depend significantly on what we help ourselves to at the outset. We should try to figure out what we think we're entitled to claim for ourselves at the outset.

Do we want to emphasize the intellectual virtues that cluster around evidential security? If so, empiricism is a good way to do this. We should stick pretty closely to what's presented in experience, and we shouldn't presume that too much is presented in experience. If we do this, we're highly likely to avoid certain mistakes. So, skepticism about induction, of the sort that Karl Popper and David Hume put forward and defend, and a resistance to Bayesian subjective probabilities (a resistance to the idea that we should think the world takes our biases, our categories, our conceptions, seriously) is a good way to avoid error.

We should admit, however, that this increased security comes at the cost of significantly diminished resources. We've seen thinkers from a number of approaches claim that science doesn't need any more than they're willing to permit. The positivists think that science doesn't need to talk seriously about unobservable reality, for instance. Maybe, but most of us think, in our heart of hearts, that there is an unobservable reality, and it costs something to remain silent and skeptical about the nature of that unobservable reality. It prevents us from saying things we might unreflectively have thought we were entitled to say (about copper *needing* to conduct electricity, not merely *happening* to conduct electricity, for instance).

So, on the other hand, you can pursue your explanatory ambitions. You can take your models seriously and think that they can describe a world you can't see. You can take your subjective probability assignments seriously, and think that you can update them in a way so that they deserve respectful consideration. You can then set out to maximize confidence and understanding. But the risks here are muddle-headedness about what you mean and mistakes about what you believe.

So, these are the vices and virtues not just of science, but of everyday life as well. Depth and breadth in our understanding of the world is an inherently risky enterprise. It has costs in terms of how clearly we understand the claims we make as we go beyond the evidence, and how strongly we can support the claims we make when we go beyond the evidence.

Another difficult tension that has arisen within this course: Kuhnian fidelity to actual science is an attractive idea, but the "is/ought" distinction is an attractive idea, and these two sit somewhat awkwardly together. We should talk about the extent to which they can be reconciled.

The smart money, I think, if you have to bet, is on scientific practice over philosophical advice about scientific practice. They've done pretty well; we haven't done quite so well. But what scientists say they're doing doesn't always reflect what, in fact, they're doing. Scientists tend to commit philosophy when they describe what they do. We've seen that many of them think, for instance, that they're

Popperians when they probably are not. Scientists don't like to admit that they've got philosophical views, but they virtually always do.

So, to some extent, they're on our turf. And anyway, in order to describe what scientists are doing in the first place, we have to bring some conception of what it is to do science to the table. We don't mimic scientists when they're going to lunch. We have some conception of which part of what they're doing counts as the part we care about, counts as doing science.

In any case, the smart money doesn't always win in any particular case. Even though, by and large, you should bet on science, sometimes there is room for informed criticism of actual scientific practice. I suggested a couple of lectures ago that—at least sometimes—the ways parts of science use statistical significance tests seems to me a little bit flat-footed (and I don't feel obligated to defer to the people in the white lab coats merely because they've got the white lab coats).

Talk about objectivity in science tends to bring out the worst in everybody. I've helped myself to this notion, with some misgivings, in this course because it's just hard to do without. But it's an intrinsically dangerous notion, I think. It tends to lead to exaggerating the virtues and/or exaggerating the vices of science and scientists.

So, people sympathetic to views like social constructivism or postmodernism hear a term like "objectivity," and they picture scientists claiming to view nature from nowhere—to step out of their skins, to step out of language, to step out of history, and to carve nature at its joints via some purely passive, cognitive process. They rightly regard most of this as pretty naïve, but that stems from a ridiculously demanding notion of objectivity being deployed.

The reason many of these people want to deflate science stems from, I think, an admirable desire to tolerate disagreement. They don't want to say that other views are wrong; it seems disrespectful to go around pronouncing that other epistemic practices are misguided. I think this embodies a serious confusion that is somewhat rampant in our culture. It is important not to confuse toleration with a lowering of epistemic standards.

A quick story from the Counter-Reformation will, I think, bring out the point nicely. There's an English philosopher of the time who made a stunning realization that he reported in a letter to his friend. The Catholics were claiming that the Protestants were all going to hell, and the Protestants were claiming that the Catholics were all going to hell. This guy realized (he was a Protestant, of course, being English at the time) that the Catholic arguments from the Catholic standpoint looked every bit as good as the Protestant arguments looked from the Protestant standpoint. He realized these were not crazy people; they were perfectly rationally reasoning from their own premises.

Does that lead to a stunning display of tolerance? No. What the guy realized is, since their arguments look reasonable to them, we can't reason with these people—we have to kill them.

There's a big difference between being tolerant and deciding that there's no fact of the matter about who's right. This guy decided that there was no way rationally to convince the other side that they were wrong. That doesn't automatically lead to tolerance, nor does thinking that people who disagree with you (when you've got, say, a well-tested, scientific theory) are wrong. That doesn't mean you have to be intolerant or a jerk to them.

This way of trying to defend toleration leads to relativism, to the view that says, to a first approximation, it's objectively true that nothing's objectively true. That's a problem. We want to defend our toleration on its own terms without muddying the epistemic waters by trying to say that there's no way that science can be better than any other intellectual enterprise.

The vices on the other side are just about equally dangerous. People sympathetic to realism or to empiricism hear other people scoffing at the idea of objectivity, and they get worked up because they think science is being reduced to rhetoric, or literature, or something like that—and they, then, want to valorize science. So, they dismiss legitimate questions about whether science is, itself, informed by values, whether there are any limits to what science can show us.

This leads to equally soft-headed thinking—the very kind of soft-headed thinking that these supposedly hard-nosed thinkers were accusing the postmodernists and constructivists of making. They start thinking that the problem of demarcation is no problem at all, and that they know what a real science is, and real science clearly gets the world right in an unproblematic way—and so they stop worrying about, say, the underdetermination of theory by data; they stop worrying about the problem of induction. That's just about equally naïve.

As this dialectical process continues, each side tends to move towards more of a caricature of itself. This is what happened in the Science Wars of the '90s. This was the most publicized—but also the ugliest—display of thinking about science in recent decades.

This is an important tension to resolve. There are many notions worth calling "objectivity" between these caricatures. Science does not just have the world imprinted on, as it were, its collective mind, but nor is it just a literary genre. Hopefully, some of these intermediate conceptions of objectivity have come out in the abstract as we started looking at various theories of evidence. The ways in which Kuhnian normal science is constructed brings with it a certain kind of objectivity that gets built out of some of these subjectivities and idiosyncrasies of individual scientists. Our thinking about probability and statistics allows for reasonably refined views about what constitutes "objectively versus subjectively" settling such questions.

Also, as we've used some examples from particular sciences, hopefully, we have exemplars—in the Kuhnian sense—of how questions can get settled in a way that's not dogmatic or naïve about science having direct access to reality, but nor is it excessively cynical in thinking that we've just constructed thermodynamics or quantum mechanics to answer to political or other idiosyncratic needs.

It's highly important, I think, for us to have a realistic sense of the place of science in our culture. Science deserves our respect; it should not be leveled down to just another text, or just an interpretation, or just an expression of the interests of those who have certain powers. But it does not deserve automatic deference. Philosophy, in general, is supposed to provide a kind of manual for intellectual self-defense. So, philosophy of science should help us look at claims made within science, and claims made about science, and can help us make informed judgments about how and what we're to think about each case.

As I said at the outset of this lecture, philosophy is hard. Science is hard. Philosophy of science is doubly hard, which is part of why those of you who have made it this far, honestly, genuinely have my admiration.

What do we get to show for all of this hard work? I haven't tried to answer many of these questions. I hope we've come away more with clarity than with knowledge. We've come away knowing that there's a balance of virtues to be struck, which is frustrating because we don't know exactly how to strike it, but clarity is liberating. We become able to see things and to say things that we couldn't have seen and couldn't have said.

There's this emotional picture about logic and rigor, according to which it's supposed to weigh us down. It's supposed to be a sort of burden—a standard we struggle to meet, as it were, for its own sake. But Bertrand Russell, I think, had it right when he suggests that, drawing clear distinctions, logical rigor is freeing because it shows us possibilities that have been there all along, but that muddle-headedness prevents us from seeing.

So, what I hope you'll take from this course is an ability to see and to think for yourself about the intellectual accomplishments of science and the intellectual virtues of science, so that you can appreciate what science offers to us, and deploy these intellectual virtues in your own life.

I'd like to thank anyone who has made it through this, I think, difficult and demanding course, and the people here at The Teaching Company who have helped me deliver it.

Timeline

6th century B.C.E. ....Thales asserts that water is the "primary principle." This is arguably the first attempt at scientific explanation and at a scientific reduction.

4th century B.C.E. ....Aristotle develops a systematic, sophisticated approach to scientific inquiry, involving both methodological and substantive advances.

c. 300 B.C.E. ....Euclid develops the standard presentation of geometry, which stood as a model of scientific perfection for 2,000 years.

c. 400 C.E. ....Evidence of sophisticated reasoning about probability appears in the Indian epic *Mahabharata*.

1543 ....Nicholas Copernicus puts forward the first detailed proposal that the Earth is a planet orbiting the Sun. The work was published with a preface by Andreas Osiander indicating that the theory should be treated as a calculating tool, not as a description of reality.

1583–1632 ....Galileo Galilei argues for the literal truth of the Copernican system, formulates a law of falling bodies and a law governing the motion of pendulums, applies the telescope to celestial phenomena, articulates a principle of the relativity of inertial motion, and generally develops a quantitative and observational approach to motion.

1605–1627	Francis Bacon develops the first systematic inductive method, a plan for attaining and increasing knowledge on the basis of experience.
1609	Johannes Kepler formulates his first two laws of planetary motion (the third law would have to wait 10 years).
1628	William Harvey establishes the circulation of the blood and the heart's function as a pump.
1633–1644	Rene Descartes invents analytical geometry and develops his highly influential physics.
c. 1660	The basic mathematics of probability takes shape in the work of Blaise Pascal, Christian Huygens, and others.
1660	The Royal Society of London for the Improving of Natural Knowledge is founded. Early members of the Royal Society include Robert Boyle, Christopher Wren, Robert Hooke, John Locke, and Isaac Newton. The Royal Society agitates in favor of experimental knowledge and against scholasticism and tradition. Many members are particularly interested in observational knowledge of witchcraft.
1661–1662	Robert Boyle takes major steps toward the separation of chemistry from alchemy, and he determines that the pressure and volume of a gas are inversely proportional. Boyle's "corpuscularian"

conception of matter greatly influenced John Locke.

1666 .... By this time, Isaac Newton had developed the fundamental principles of calculus, had formulated the principle of universal gravitation, and had established that white light consists of light of all colors of the spectrum.

1673 .... Molière, in his play *Le malade imaginaire*, makes fun of the explanation that opium puts people to sleep because it has a "dormitive virtue."

1678 .... Christian Huygens puts forward a version of the wave theory of light.

1687 .... Isaac Newton publishes his monumental *Principia*, which contains all the basic features of his mechanics, including his explicitly absolute conception of space and time.

1690 .... John Locke publishes his masterpiece, *An Essay Concerning Human Understanding*.

1704 .... Isaac Newton defends the particle (or corpuscular) theory of light in his *Opticks*.

1709, 1714 .... Gabriel Daniel Fahrenheit constructs an alcohol thermometer and, five years later, a mercury thermometer.

1710 .... Publication of George Berkeley's most important work, *A Treatise Concerning the Principles of Human Knowledge*.

1715–1716 ....Gottfried Wilhelm Leibniz develops a sophisticated relational account of space and time through a critique of Newton's work.

1738 ....Daniel Bernoulli publishes an early version of the kinetic theory of gases.

1739–1740 ....David Hume publishes his most important work, *A Treatise of Human Nature*.

1750s..Carl von Linne (also known as Carolus Linneaus) launches the modern taxonomic system involving genera and species.

1751 ....Benjamin Franklin publishes *Experiments and Observations on Electricity*.

1763 ....Thomas Bayes's paper containing his famous theorem is presented to the Royal Society by Bayes's friend Richard Price.

1769 ....James Watt patents his steam engine.

1770s..Joseph Priestly isolates a number of gases, including "dephlogisticated air," soon to be renamed *oxygen* by Antoine Lavoisier.

1777 ....Lavoisier performs the experiments that doom the phlogiston theory of combustion.

1789 ....Lavoisier establishes that mass is conserved in chemical reactions and formulates the modern distinction between chemical elements and compounds.

1795	James Hutton publishes *Theory of the Earth*, considered by many to be the founding document of the science of geology.
1808	John Dalton's *New System of Chemical Philosophy* propounds the atomic theory of chemistry.
1809	Jean-Baptiste Monet de Lamarck proposes the first really significant theory of evolution. Lamarck emphasizes the heritability of acquired characteristics.
1818	Simeon Poisson deduces from Augustin Fresnel's wave theory of light the apparently absurd consequence that a bright spot will appear at the center of the shadow of a circular object under certain conditions. Dominique Arago almost immediately verifies the prediction, however.
1824	Nicolas Léonard Sadi Carnot, despite relying on a conception of heat as a kind of substance, works out many of the central ideas of thermodynamics.
1826	Nikolai Ivanovich Lobachevsky produces a geometry that replaces Euclid's Fifth Postulate and allows more than one line parallel to a given line to pass through a fixed point.
1830	August Comte distinguishes theological, metaphysical, and positive stages of history, giving currency to the term *positivism*.

1832	Poisson proves a version of the law of large numbers and offers the clearest distinction yet drawn between "relative-frequency" and "degree-of-belief" approaches to probability.
1840	William Whewell develops a conception of scientific methodology that is hypothetical rather than purely inductive. In the same work, Whewell introduces the term *scientist* into the English language.
1844	Adolphe Quetelet argues that the bell-shaped curve that had been applied to games of chance and to astronomical errors could also apply to human behavior (for example, to the number of murders in France per year).
1850	Rudolph Julius Emanuel Clausius, generalizing Carnot's work, introduces a version of the second law of thermodynamics.
1859	Charles Darwin publishes his epoch-making *On the Origin of Species*.
1861	James Clerk Maxwell reduces light to electromagnetic radiation.
1866	Gregor Mendel develops his theory of heredity involving dominant and recessive traits.
1869	Dmitri Ivanovich Mendeléev develops his periodic table of the elements.
1870s	Ludwig Boltzmann offers two different reconciliations of the time directionality of the laws of

	thermodynamics with the time reversibility of the basic laws of motion.
1878	In Leipzig, Wilhelm Wundt establishes the first laboratory for physiological psychology.
1879	Gottlob Frege publishes his *Begriffsschrift*, arguably the founding document of modern mathematical logic.
1887	A. A. Michelson and E. W. Morley measure the speed of light as the same in all directions and thereby fail to detect any motion of the Earth with respect to the aether.
1889, 1892	G. F. Fitzgerald and H. Lorentz independently suggest that the null results of the Michelson-Morley experiments can be explained on the assumption that physical objects (such as measuring devices) contract at speeds approaching that of light.
1892	C. S. Peirce argues that there is no compelling scientific or philosophical reason for accepting determinism.
1895	X-rays are discovered by W. C. Röngten.
1900	Max Planck introduces the "quantum theory," according to which light and energy are absorbed and emitted only in bundles, rather than continuously.
1902	Ivan Pavlov carries out his well-known experiments involving learning and conditioned responses.

1905	Bertrand Russell publishes "On Denoting," which becomes a paradigm of philosophical analysis.
1905	Albert Einstein publishes enormously important papers that, among other things, formulate the special theory of relativity and help explain Planck's quantum theory.
1912	John Watson advocates behaviorism as the scientifically appropriate approach to psychology.
1912	A. L. Wegener proposes a unified theory of continental drift.
1913	Niels Bohr publishes "On the Constitution of Atoms and Molecules," which is often taken to contain the first theory of quantum mechanics.
1915	Einstein publishes his general theory of relativity.
1919	A team led by Arthur Eddington obtains experimental confirmation of Einstein's hypothesis that starlight is bent by the gravitational pull of the Sun.
1923	Louis Victor de Broglie suggests that the wave-particle duality applies to matter as well as to light.
1926	Frank Ramsey, in "Truth and Probability," lays much of the foundation for a rigorous interpretation of probabilities as degrees of belief.
1926	Max Born interprets electron waves probabilistically; the electron is more likely to be found in places

	where the square of the magnitude of the wave is large than where it is small.
1927	Werner Heisenberg denies that an electron simultaneously possesses a well-defined position and a well-defined momentum.
1927	Percy Bridgman's "The Operational Character of Scientific Concepts" is published.
1927	Bohr and others formulate the Copenhagen interpretation of quantum mechanics.
1929	Edwin Hubble observes that all galaxies are moving away from one another.
1929	Rudolf Carnap, Otto Neurath, and Hans Hahn publish "The Vienna Circle: Its Scientific Outlook," a manifesto of logical positivism.
1931	Sewall Wright argues that random genetic drift plays a significant role in evolution.
1931	Kurt Gödel's incompleteness proof is published. This shows that any axiomatic system powerful enough to include arithmetic will imply at least one provably false consequence.
1934	Karl Popper publishes *The Logic of Scientific Discovery*.
1936	The first edition of *Language, Truth and Logic*, by A. J. Ayer, appears.
1942	Ernst Mayr publishes *Systematics and the Origin of Species*, a

watershed work in biological classification.

1942 .... Julian Huxley publishes *Evolution: The Modern Synthesis*, which unified many aspects of biological research that had been achieved through the work of R. A. Fisher, Sewall Wright, J. B. S. Haldane, and others.

1945 .... Carl Hempel's "Studies in the Logic of Confirmation," which includes the raven paradox, appears in print.

1948 .... Hempel and Paul Oppenheim publish the first major statement of the covering-law theory of explanation.

1951 .... W. V. Quine publishes "Two Dogmas of Empiricism." It appears in book form in 1953.

1953 .... James Watson and Francis Crick ascertain the chemical structure of DNA.

1954 .... Nelson Goodman's *Fact, Fiction and Forecast*, which includes the classic statement of the new riddle of induction, appears.

1961 .... Ernest Nagel's *The Structure of Science*, which presents a sophisticated positivist conception of science and includes a classic account of scientific reduction, is published.

1962 .... *The Structure of Scientific Revolutions*, by Thomas Kuhn, appears in print.

1963	J. J. C. Smart famously argues, in *Philosophy and Scientific Realism*, that there are no laws in biology. Smart's work is also sometimes taken to mark the resurgence of interest in scientific realism.
1963	Murray Gell-Mann and George Zweig independently arrive at the notion of quarks. Zweig treats them as tiny particles, while Gell-Mann thinks of them more as patterns than as objects.
1969	Quine's essay "Natural Kinds," a landmark of philosophical naturalism, is published.
1970	Saul Kripke presents the causal (also known as historical chain) theory of reference in lectures that would eventually be published as *Naming and Necessity*.
1970–1971	The most important papers outlining Imre Lakatos's methodology of scientific research programs appear.
1972	Stephen Jay Gould and Niles Eldredge argue that evolution largely proceeds in fits and starts, rather than gradually. This is known as the *punctuated equilibrium* approach to evolution.
1974	Michael Friedman publishes an influential account of explanation as unification.
1974	*The Structure of Scientific Theories*, a volume edited by Frederick Suppe, appears in print. The book contains classic presentations of the "received view" of scientific

	theories and of the then-new semantic conception of theories.
1974	*Knowledge and Social Imagery*, a classic work in the *strong program* in the sociology of knowledge, is published by David Bloor.
1975	Paul Feyerabend's *Against Method* appears in print.
1977	Larry Laudan's *Progress and Its Problems*, which includes a classic statement of the pessimistic induction argument against scientific realism, is published.
1980	Bas van Fraassen publishes *The Scientific Image*, which details both his constructive empiricism and his approach to explanation.
1981	Paul Churchland defends an influential version of eliminative materialism, the view that folk psychology is radically false and will be replaced.
1982	An Arkansas judge decides, in *McLean v. Arkansas Board of Education*, that creation-science does not count as science. The case included testimony about the problem of demarcation and has occasioned a great deal of discussion.
1983	David Armstrong's *What Is a Law of Nature?* awakens interest in non-regularity accounts of physical laws.
1985	Steven Shapin and Simon Schaffer's *Leviathan and the Air-Pump*, an important work in historical

	sociology of knowledge, is published.
1988	David Hull's *Science as a Process*, which examines such matters as the social structure and reward system of science, is published.
1990	Helen Longino publishes *Science as Social Knowledge*, a major work concerning social structure and objectivity.
1990	Philip Kitcher's influential work on the division of cognitive labor appears in *The Journal of Philosophy*.
1996	Alan Sokal's parody of postmodernism, "Transgressing the Boundaries: Toward a Transformative Hermeneutics of Quantum Gravity," is published in *Social Text*, and Sokal reveals his hoax in *Lingua Franca*. This period sees the height of the so-called Science Wars.

Glossary

analytic/synthetic: Analytic statements have their truth or falsity determined by the meanings of the terms of which they are composed. "No triangle has four sides" is an example of a (supposed) analytic truth. The truth value of synthetic statements, such as "All copper conducts electricity," depends, not just on what the statement means, but also on what the world is like. Quine denies that any statements are properly regarded as analytic.

a priori/a posteriori: This distinction concerns how the truth or falsity of a statement can come to be known. A statement is knowable *a priori* if the justification of the statement does not depend on experience. You may, in fact, have learned that $2 + 2 = 4$ through experience (counting apples and oranges and such), but if the justification for this claim is not experiential (if, for instance, the claim is analytically true), then it is knowable *a priori*. Statements not knowable *a priori* are knowable only *a posteriori*, that is, in part on the basis of evidence obtained through experience (though not necessarily one's own experience).

auxiliary hypotheses: We generally have a sense of which hypothesis we mean to be testing. But no hypothesis has any observational implications all by itself; thus, we must include auxiliary hypotheses in order to derive predictions from the hypothesis under test. Even if the predictions prove false, it is possible that the hypothesis under test is true and that the false prediction should be "blamed" on one of the auxiliary hypotheses.

Bayesianism: Although a range of probability-centered approaches to the theory of evidence and confirmation can be considered Bayesian, orthodox Bayesians interpret probability statements as degrees of belief, and they permit a great deal of "subjectivity" in the assignment of prior probabilities. They require that one update one's degrees of belief in accordance with Bayes's Theorem.

bridge law: Bridge laws are crucial to scientific reductions, at least as classically understood. If, as is generally the case, the theory to be reduced contains terms that do not appear in the reducing theory, bridge laws are used to connect the vocabulary of the two theories. "Temperature is mean molecular kinetic energy" is a rough statement of a classic bridge law. Without bridge laws, the reduced theory cannot be logically derived from the reducing theory.

causal model (of explanation): The main successor to the covering-law model of explanation, the causal model says that events are explained by revealing their causes.

cognitive meaning: Cognitively meaningful statements are literally true or false. They are contrasted with sentences that don't aspire to make true assertions (such as questions, commands, and poetry) and, more important for our purposes, with metaphysical statements that, according to the logical positivists, aspire to cognitive meaningfulness but fail to achieve it.

concept empiricism: This position, exemplified by Hume, asserts that any legitimate concept must be traced back to sources in direct experience. Concepts that cannot be so traced (for example, substance) are not genuinely meaningful.

constructive empiricism: Bas van Fraassen's constructive empiricism combines an empiricist conception of evidence (according to which all evidence is observational evidence, and the distinction between the observable and the unobservable is of great importance) with an anti-empiricist conception of meaning (scientific theories refer to unobservable reality in much the same way that they refer to observable reality). As a result, van Fraassen maintains that good theories are committed to claims about unobservables but that good scientists need not believe what their theories say about unobservables.

constructivism: This term is sometimes rendered *constructionism*. Generally, this is the idea that (some part of) reality is made rather than found. In the context of our course, this idea gets its start with Kuhn's suggestion that paradigms help determine a scientist's world or reality. Constructivists tend to be suspicious of the distinction between experience, theories, and beliefs, on the one hand, and reality, on the other.

context of discovery: The empiricisms of the 17th–19th centuries tried to formulate rules that would lead to the discovery of correct hypotheses. The dominant empiricist views of the 20th century relegated discovery to psychology and sociology; they held that scientific rationality applies only in the context of justification. The distinction between discovery and justification has been under pressure since Kuhn's work became influential.

context of justification: The positivists and other 20th-century empiricists held that, although no method for generating promising hypotheses is available, once a hypothesis has been generated, a logic or method can be found by which its justification can be assessed. They thus held that, although there is no logic of scientific discovery, there is one of scientific justification.

contingent/necessary: A contingently true statement is actually true (another way of saying this is to call the statement "true in the actual world") but could be false (that is, the statement is false in some possible world). Necessary truths hold in all possible worlds. In certain contexts, particular kinds of necessity or contingency are at work. For instance, it is physically necessary (that is, necessary given the laws of nature) that copper conduct electricity, but it isn't logically necessary that it do so (that is, there is no contradiction involved in the idea of nonconductive copper).

corroboration: This is Popper's term for theories or hypotheses that have survived serious attempts to refute them. Because Popper insists that corroboration has nothing to do with confirmation, he claims that we have no reason to think corroborated theories more likely to be true than untested ones.

counterfactual: Counterfactual conditionals are expressed in the subjunctive, rather than in the indicative mood. "If my coffee cup were made of copper, it would conduct electricity" is a counterfactual conditional. Counterfactuals can be used to test how robust a statement is, that is, how insensitive the truth of the statement is to actual circumstances.

covering-law model: The centerpiece of logical positivism's philosophy of explanation, the covering-law model treats explanation as the derivation of the explanandum from an argument containing at least one law of nature.

deduction/deductive logic: This is the relatively unproblematic, well-understood part of logic. It is concerned with the preservation of truth. If an argument is deductively valid, then it is impossible for the premises of the argument to be true while the conclusion is false.

demarcation criterion: A demarcation criterion would provide a basis for distinguishing science from pseudoscience.

determinism: Determinism holds if the state of the universe at a given moment suffices to exclude all outcomes except one. Generally, determinism is understood as causal determinism: the state of the universe at a given moment causally determines the outcome at the next moment. Quantum mechanics suggests that the universe is not deterministic.

disposition: Dispositions manifest themselves only under certain conditions. A substance is soluble (in water) if it is disposed to dissolve when placed in water. Because substances are taken to retain their dispositional properties even when they are not in the relevant circumstances, dispositional properties outrun their manifestations in experience and, thus, pose problems for empiricists.

eliminative reduction: Generally, when some "stuff" (water) or a theory (thermodynamics) is reduced to something else (H_2O or statistical mechanics), the reduced entity or theory does not lose any of its claim to real existence. Sometimes, though, the right sort of reduction eliminates the existence of the thing reduced. When we reduce cases of demonic possession to certain kinds of illness, we thereby show that there were never any cases of demonic possession.

empiricism: A wide range of views can lay claim to this label. They all have in common some conception, according to which experience is the source of some cognitive good (for example, evidence, meaningfulness). See **concept empiricism** and **evidence empiricism**.

entailment: Statement A entails statement B if it is impossible for A to be true without B being true.

epistemology: A fancy Greek word meaning the theory of knowledge and justification.

evidence empiricism: This is the thesis that all of our evidence (at least all of our evidence for synthetic propositions) ultimately derives from experience. Rationalists, in contrast, think that some synthetic statements can be justified on the basis of reason alone.

exemplar: An exemplar is a Kuhnian paradigm in the narrow sense of that term. Exemplars are model solutions to scientific puzzles. Exemplars loom very large in scientific education, according to Kuhn.

explanandum: A fancy Latin word meaning "that which is explained." It often refers to a *sentence* describing the event (or whatever) being explained.

explanans: A fancy Latin word meaning "that which is explaining."

explanatory inference: See **inference to the best explanation**.

falsificationism: Popper's demarcation criterion and his conception of scientific testing are generally combined under this term. Science is distinguished from pseudoscience by the readiness with which scientific claims can be falsified. In addition, scientific testing can falsify but can never confirm theories or hypotheses.

folk psychology: It is much disputed whether folk psychology is a theory or not. We explain one another's behavior in terms of beliefs, desires, and so on, and this explanatory and predictive practice is folk psychology, whether it merits being considered a psychological theory or not.

functional properties: Some objects are individuated by what they do (or what they're for), rather than by what they're made of. Knives can be made of any number of materials; they are united by their purpose or function.

holism: In the context of this course, holism is associated with the work of Quine, who emphasizes holism about testing—no hypothesis can be tested without extensive reliance on auxiliary hypotheses—and holism about meaning—because, in the positivist tradition, testability and meaning are closely linked, statements and terms are meaningful only in the context of a whole theory.

Hume's fork: Hume's fork is basically a challenge grounded in his empiricism. All meaningful statements concern either "matters of fact" and are subject to the empirical sciences or "relations of ideas" and are, at bottom, analytic and the proper domain of such disciplines as mathematics. It is a matter of some controversy whether Hume's fork leaves any space for philosophy.

hypothetico-deductive: Another bit of pure poetry, brought to you courtesy of philosophers of science. The hypothetico-deductive model of confirmation is simple and powerful. It says that a hypothesis is confirmed when true observational consequences can be deduced from it. If the hypothesis (along with auxiliary hypotheses, of course) makes observational predictions that turn out to be false, then the hypothesis is disconfirmed or, perhaps, even refuted.

incommensurability: Literally, this term refers to the lack of a common measure. In the work of Kuhn, Feyerabend, and their successors, incommensurability indicates a range of ways in which competing paradigms resist straightforward comparison. Insofar as two paradigms offer different standards for scientific work and assign different meanings to crucial terms, it will be difficult to assess them in terms of plausibility, promise, and so on.

induction: There is little agreement about how this term should be used. In the narrow sense, induction comprises "more-of-the-same" inferences. A pattern is carried forward to new cases. Some thinkers would assimilate analogical inference to this pattern. In the broad sense, induction includes explanatory inferences, as well as analogical and "more-of-the-same" inferences.

inference to the best explanation: This encompasses a range of inferential practices (such terms as *abductive inference* and *explanatory inference* are sometimes used to mark differences within this range). The general idea is that a theory's explanatory success provides evidence that the theory is true. This style of argument is crucial to scientific realism but is regarded with some suspicion by empiricists.

instrumentalism: Sometimes, any version of anti-realism about science is called instrumentalist, but it is probably more useful to reserve the term for the idea that scientific theories are tools for predicting observations and, thus, do not have to be true to be good (though they have to lead to true predictions in order to be good).

laws of nature: Not all laws of nature are called laws. Some fundamental and explanatory statements within sciences are called equations, for instance. The philosophical disagreement (between regularity theorists and necessity theorists, mainly) concerns what makes a true, fundamental, and explanatory statement a law of nature.

logical empiricism: See **logical positivism**.

logical positivism: In this course, *logical positivism* and *logical empiricism* are used interchangeably. These terms refer to an ambitious, language-centered version of empiricism that arose in Vienna and Berlin and became the standard view in philosophy of science through the middle of the 20th century. Under the pressure of criticism (largely from within), the positivist program became somewhat more moderate over the years.

metaphysics: This term was generally used pejoratively by the positivists to refer to unscientific inquiries into the nature of reality. These days, most philosophers see room for a philosophical discipline worth calling metaphysics, which addresses such issues as personal identity, the reality of universals, and the nature of causation.

model: Models can be abstract or concrete. In either case, the structure of the model is used to represent the structure of a scientific theory. This semantic approach to theories contrasts with the syntactic approach characteristic of positivism and the "received view" of scientific theories.

naturalism: Naturalism has been enormously influential in recent philosophy. It comes in many flavors, but the central ideas include a modesty about the enterprise of philosophical justification and a consequent emphasis on the continuity between philosophy and science. Naturalists give up on the project of justifying science from the ground up and thereby free themselves to use scientific results (for example, about how perception works) for philosophical purposes.

natural kinds: The contrast is, unsurprisingly, with artificial kinds. The notion of a natural kind can receive stronger and weaker construals. Strongly understood, natural kinds are nature's joints, grouping things that are objectively similar to one another. Chemical elements might be thought of this way; biological species are a harder case. More weakly, natural kinds are the categories that matter to scientific theorizing.

necessary: See **contingent/necessary**.

necessary condition: A is a necessary condition for B just in case nothing can be B without being A. Being a mammal is a necessary condition for being a whale.

necessitarian view of laws: Unlike regularity theorists, necessitarians maintain that laws of nature do more than just report what invariably happens. Necessitarians think that the laws of nature report relations among universals or similar "deep" features that *make*, for example, copper conduct electricity.

normative: This term contrasts with *descriptive*. Normative claims concern how things ought to be rather than how they are.

objective: A term that probably does more harm than good, but one that is nevertheless nearly impossible to avoid. *Objective* can modify such things as beliefs, in which case it refers to the absence of bias or idiosyncrasy. It can also modify such a term as *existence*, in which case it indicates that something exists independently of its being thought of, believed in, and so on.

ontology: In philosophy, ontology is the part of metaphysics concerned with existence. The ontology of a scientific theory is the "stuff" (objects, properties, and so on) that, according to the theory, exists.

operationalism: Also sometimes called *operationism*, this influential approach to the meaning of scientific terms originated with the physicist P. W. Bridgman. It requires that scientific terms be defined in terms of operations of measurement and detection. This approach is generally thought to be too restrictive.

paradigm: For the narrow use of paradigm, see **exemplar**. In the broad sense, a paradigm includes exemplars but also theories, standards, metaphysical pictures, methods, and whatever else is constitutive of a particular approach to doing science.

partial interpretation: This term contrasts, unsurprisingly, with *full interpretation*. Because the positivists held that meaning arises from experience, they had a difficult time assigning full meaning to statements that go beyond experience. They minimized this problem with their idea of theories as partially interpreted systems. Even if a term such as *fragility* can be applied only to objects that meet certain test conditions, the term is still useful for generating predictions and for connecting observations to one another.

pessimistic induction: This refers to one of the major arguments against scientific realism. Most successful scientific theories have turned out to be false, so we should expect that currently successful theories will turn out to be false.

positivism: In this course, *positivism* is generally used as an abbreviation for *logical positivism*. The term also refers to a 19th-century version of empiricism associated with August Comte. Comte defended a more extreme version of empiricism than did "our" positivists. For instance, he denied that science aspires to explain phenomena.

posterior probability: This is the probability of a hypothesis given some evidence. It is represented as P(H/E) or as P(H/E&B) if we want to make the role of background evidence explicit. P(H/E) is usually spoken as "the probability of H on E" or "the probability of H given E."

prior probability: This can either mean the probability of a hypothesis before any evidence at all has been gathered or the probability of the hypothesis before a particular piece of evidence is in. Either way, the prior probability is written P(H).

probability: A mathematical notion, but one that can receive a range of interpretations. We are mainly concerned with Bayesians, for whom probabilities are understood as degrees of belief. Others understand probability statements in terms of (actual or idealized) frequencies or physical propensities, among other possibilities.

problem of old evidence: It would seem that any evidence we already know to be true should receive a probability of 1. But if we plug that value into Bayes's Theorem, we can see that any evidence that has a probability of 1 cannot confirm any hypothesis in the slightest.

rational reconstruction: Popper and the positivists tended to offer rational reconstructions of scientific practice. A rational reconstruction characterizes the justified core of a practice, rather than the practice as a whole. Largely as a result of Kuhn's work, philosophers have been less confident in recent years that they can isolate the rational core of science.

realism: See **scientific realism**.

"received view" of theories: See **syntactic conception of theories**.

reduction: A reduction occurs when a more general theory can account for the (approximate) truth of a more specific theory. The standard or classical account of reduction favored by the positivists requires the reduced theory to be derivable from the reducing theory plus suitable bridge laws. Reduction, insofar as it happens, appears to offer an unproblematic sense in which science makes progress.

regularity view of laws: Regularity theorists maintain that laws of nature comprise a subset of nature's regularities, namely, things that always happen. Laws do not involve any kind of causal necessity, as they do on the rival necessitarian conception of laws.

relativism: The kind of relativism at issue in this course concerns justification or truth. A relativist denies that standards of justification or truth can be applied independently of such things as theories, paradigms, or class interests. The standards are then said to be relative to the theories or interests. Objectivists about justification or truth think that we can make useful sense of these notions independently of our theories or interests.

research program: Lakatos's notion of a research program is loosely analogous to Kuhn's notion of a paradigm (in the broad sense). Lakatos allows competition among research programs and imposes a more definite structure on his research programs than Kuhn had on his paradigms. Research programs involve a *hard core* of claims that are not subject to test and a *protective belt* of claims that can be modified in the light of experience.

scientific realism: Another idea that comes in several flavors, scientific realism has at its core the claims that scientific theories aim to correctly depict both unobservable and observable reality and that, in general at least, adopting a scientific theory involves believing what it says about all of reality.

scientific revolution: This term was made famous by Kuhn, but one needn't be a Kuhnian to think that Newton, Darwin, and Einstein, among others, revolutionized science. Kuhn is skeptical about whether traditional notions of progress and accumulation hold across revolutions, but any view of the history of science will have to make some sense of the enormous changes to scientific practice that have occasionally taken place.

semantic conception of theories: Against the received, or syntactic, view of theories, the semantic approach treats theories as sets of models rather than as axiomatic systems. The semantic approach does not rely as heavily as does the syntactic on the distinction between observable and unobservable reality.

strong program: The strong program is an influential approach within the sociology of science. It seeks to explain scientific behavior by examining the psychological and sociological causes of beliefs and decisions. The strong program's most controversial component is the symmetry principle, according to which the truth or justification of a belief should play no role in explaining its acceptance.

sufficient condition: A is a sufficient condition for B just in case anything that is A must be B. Being made of copper is sufficient for being metallic.

supervenience: To say that one domain supervenes on another is to say that there can be no change at the "upper" level without a change at the "lower" level. For instance, to say that the domain of the psychological supervenes on the domain of the physical is to say that any two situations that are physically identical would have to be psychologically identical.

syntactic conception of theories: Also known as the received view of theories, this approach conceives theories as systems of sentences modeled, more or less, on geometry. The fundamental laws of the theory are the unproved axioms. In its classic, positivist incarnations, meaning "flows up" into the theory from observation statements, and a theory is, thus, a "partially interpreted formal system."

synthetic: See **analytic/synthetic**.

teleological explanation: An explanation that makes reference to a purpose is said to be teleological. Such explanations are prevalent in biology (creatures have hearts for the purpose of pumping blood) and psychology (we explain behavior as goal-directed). Philosophers have worked hard to reconcile teleological explanation with non-purposive explanation.

theory-ladenness of observation: Another position that comes in various strengths, claims about theory-ladenness range from uncontroversially modest ones (for example, that the theory one holds will affect how observations are described) to highly controversial ones (notably that observation cannot provide any sort of neutral evidence for deciding between theories or paradigms).

underdetermination: This is generally understood to be shorthand for "underdetermination of theory by evidence." This thesis is particularly associated with Quine's holism. For any given set of observations, more than one theory can be shown to be logically compatible with the evidence. More threatening versions of underdetermination maintain that even if additional criteria are imposed (mere logical consistency with the data is, after all, rather weak), no rational basis for settling on a theory will emerge.

unificationist models (of explanation): A recently influential approach, according to which science explains by minimizing the number of principles and argument styles we have to treat as basic. Understanding is increased when the number of unexplained explainers is minimized.

universal generalization: This is the logical form of most laws of nature. "All As are Bs" is the easiest rendering in English of this form.

verification principle: Though it has never quite received a satisfactory formulation, the verification (or verifiability) principle of meaning stood at the center of the logical positivist program. It asserts, roughly, that the meaning of any empirical statement is the method of observationally testing that statement.

Biographical Notes

Ayer, Alfred Jules (1910–1989). Ayer made a splash as a young man when he published *Language, Truth and Logic* in 1936. That book provides the classic statement in English of logical positivism. Ayer remains best known for this brash, youthful book, but he went on to do important work in several areas of philosophy. He also made a record with Lauren Bacall!

Berkeley, George (1685–1753). Berkeley was born near Kilkenny, Ireland. He became an ordained Anglican minister in 1710 and was appointed bishop of Cloyne in 1734. His first important philosophical work concerned the theory of vision, and he later incorporated results into the God-centered, immaterialist conception of the world defended in his most important book, *A Treatise Concerning the Principles of Human Knowledge*, published in 1710. Alexander Pope said that Berkeley possessed "ev'ry virtue under heav'n."

Bridgman, Percy (1882–1961). Bridgman was born in Cambridge, Massachusetts; attended Harvard; and spent his academic career there. He received the Nobel Prize for Physics in 1946 for his work on the properties of materials subjected to high pressures and temperatures. Bridgman's experimental work has proved important in geology and for such processes as the manufacture of diamonds. His most important work in the philosophy of physics is *The Logic of Modern Physics*, which was published in 1927.

Carnap, Rudolf (1891–1970). Born and educated in Germany, Carnap joined the Vienna Circle in the 1920s. In 1928, he published *The Logical Structure of the World*, an ambitious attempt to reduce talk of objects and such to experiential terms. He made a number of major contributions to philosophy in the logical positivist tradition, including crucial work on inductive logic and the structure of scientific theories. Carnap came to the United States in 1935 and spent most of his academic career at the University of Chicago and at UCLA.

Einstein, Albert (1879–1955). Nothing in Einstein's early career would have led anyone to expect his *annus mirabilis*, which took place in 1905. During that year, he published the essentials of special relativity and did groundbreaking work on Brownian motion and the photoelectric effect. He completed his general theory of relativity in 1915, and when that theory received impressive confirmation from the eclipse experiment of 1918, Einstein became an international celebrity. Einstein spent much of the rest of his scientific career pursuing a grand unified theory, and he also made important contributions to the philosophy of science, all the while crusading for peace.

Feyerabend, Paul (1924–1994). Feyerabend was born in Vienna just as the Vienna Circle was coming together. He was shot in the spine in 1945 while serving in the German army. After studying singing, history, sociology, and physics, he wrote a philosophy thesis and went to England to study with Karl Popper. Feyerabend's critiques of the dominant empiricist accounts of observation and reduction culminated in his rejection of the whole idea of scientific method, as most influentially expressed in his 1975 book, *Against Method*. Late in his career, Feyerabend spent much of his time articulating and defending philosophical relativism. Most of his academic career was spent at the University of California at Berkeley.

Goodman, Nelson (1906–1998). Goodman was born in Massachusetts and educated at Harvard. Before beginning his teaching career, he was the director of a Boston art gallery. He taught at the University of Pennsylvania and at Brandeis before joining the Harvard faculty in 1968. Though he is best known for his "new riddle of induction" (also known as the "grue problem"), Goodman made important contributions to aesthetics, philosophy of language, and epistemology, as well as philosophy of science. His 1978 book *Ways of Worldmaking* probably provides the best introduction to his distinctive approach to philosophical questions.

Hempel, Carl Gustav (1905–1997). Born in Orianenburg, Germany, Hempel studied logic, mathematics, physics, and philosophy at several German universities. He was a member of the Berlin Circle of logical positivists before moving to Vienna to work with members of the Vienna Circle. After coming to the United States in 1939, Hempel taught at Queens College, Yale University, Princeton University, and the University of Pittsburgh. Hempel's covering-law approach to explanation dominated the field for decades, and he made important contributions to the theory of confirmation, as well. His introductory text, *Philosophy of Natural Science* (1966), is regarded as a classic and offers a clear and readable approach to the field.

Hume, David (1711–1776). Often considered the greatest of the empiricist philosophers, Hume was born in Edinburgh. His *A Treatise of Human Nature*, written while Hume was in his 20s, is now regarded as one of the great works of modern philosophy but was largely ignored at the time. Doubts and whispers about Hume's religious views prevented him from ever attaining an academic position in philosophy. Hume's six-volume *History of England* did provide him with a good measure of literary success, however. His posthumously published *Dialogues Concerning Natural Religion* is generally considered a masterpiece. Hume counted Adam Smith among his good friends, and he befriended Jean-Jacques Rousseau, though they later had a very public falling out.

Kuhn, Thomas S. (1922–1996). Born in Cincinnati, Ohio, Kuhn did his undergraduate work at Harvard and received his Ph.D. in physics from the same institution in 1949. By that point, however, he had developed serious interests in the history and philosophy of science. Kuhn began his teaching career at Harvard before moving to Berkeley, Princeton, and, finally, to M.I.T. *The Structure of Scientific Revolutions* (first published in 1962) made such terms as *paradigm* and *incommensurability* part of everyday academic discourse. The book remains enormously influential. Though *Structure* issued serious challenges to the picture of science as unproblematically progressive, cumulative, and objective, Kuhn himself saw science as an unrivaled epistemic success story.

Lakatos, Imre (1922–1974). Lakatos was born and raised a Jew, though he later converted to Calvinism. His mother and grandmother died in Auschwitz. He worked in a Marxist resistance group during the Nazi occupation of his native Hungary. The communist government after the war placed him in an important position in the Ministry of Education, but he was arrested for "revisionism" in 1950. He spent almost four years in prison, including a year in solitary confinement. Lakatos took a leadership role in the Hungarian uprising of 1956 and left his native country after the Soviets suppressed the rebellion. His Ph.D. dissertation (written at Cambridge University) eventually became *Proofs and Refutations*, a remarkable work in the philosophy of mathematics. After receiving his doctorate, Lakatos joined the Popper-dominated philosophy department at the London School of Economics, where he remained until his premature death. He spent the bulk of his career developing and defending his *methodology of scientific research programs*, in which he tried to combine a Kuhnian historicism with an objective methodological standard.

Locke, John (1632–1704). One of the most influential philosophers of the modern period. Locke was educated at Westminster School in London and at Christ Church, Oxford. While at Oxford, he studied and eventually taught logic, rhetoric, and moral philosophy. He also became interested in the relatively new experimental and observational approach to medicine. In 1667, Locke moved to London as the personal physician, secretary, researcher, and friend of Lord Ashley. Lord Ashley eventually became the First Earl of Shaftesbury and Lord Chancellor, and through him, Locke became deeply involved in the turbulent politics of the period. Locke's most important work, *An Essay Concerning Human Understanding* (1690), stems in part from political motives, as Locke hoped to determine which questions could be addressed by human reason so that fruitless debates could be avoided. Other important works, including *Two Treatises of Government* (1690) and *Letter Concerning Toleration* (1690), more directly reflect Locke's concern with public life.

Mill, John Stuart (1806–1873). Mill's father, James, was himself an important philosopher, and he gave his son a remarkably intense education (Mill began reading Greek at the age of 3). His rigorous childhood left the younger Mill intellectually precocious but emotionally stunted, and he suffered a debilitating "mental crisis" in his early 20s. An exposure to the arts helped him overcome his depression. Mill is probably best known as a moral and political philosopher; *Utilitarianism* (1863), *On Liberty* (1859), and *The Subjection of Women* (1869) are classics in those fields. Mill spent much of his life working for the East India Company and was a member of Parliament from 1865 to 1868. His thoroughgoing empiricism in epistemology and metaphysics emerges in his *System of Logic* (1843) and *Examination of Sir William Hamilton's Philosophy* (1865). Mill's *Autobiography* (1873) is also a classic.

Popper, Karl (1902–1994). Popper grew up in Vienna and took his Ph.D. from the University of Vienna in 1928. He shared many scientific and philosophical interests with the members of the Vienna

Circle but disagreed with them enough that he was not invited to become a member. Popper's *Logic of Scientific Discovery* (1934) presented his own falsificationist, anti-inductive conception of scientific inquiry, along with his criticisms of logical positivism. The book remains influential to this day. The rise of Nazism forced Popper to flee to the University of Canterbury in New Zealand, where he turned his attention to social and political philosophy. *The Poverty of Historicism* (1944) and *The Open Society and Its Enemies* (1945) are products of that period. Popper moved to the London School of Economics in 1949 and was knighted in 1964.

Quine, Willard van Orman (1908–2000). Quine was born in Ohio and attended Oberlin College. He did his graduate work at Harvard and spent his academic career there. Soon after receiving his Ph.D., Quine traveled to Vienna, where he worked with the leading positivists, and to Prague, where Carnap was then living. Carnap made an enormous impression on Quine, and much of Quine's work can usefully be seen as responding to the problems faced by Carnap's versions of positivism and post-positivism. Quine's many influential papers in the philosophy of language, logic, philosophy of mind, and science are scattered through a number of collections.

van Fraassen, Bastian (1941–). Born in the Netherlands and educated at the University of Alberta and University of Pittsburgh, van Fraassen has spent much of his career working out what empiricism can amount to after the demise of positivism. He has taught at Yale University, the University of Toronto, and the University of Southern California, and he has been on the faculty at Princeton since 1982. His most influential work in the philosophy of science is *The Scientific Image* (1980), and he has also done important work in philosophical logic.

Bibliography

Essential Reading: General Anthologies

Balashov, Yuri, and Alex Rosenberg, eds. *Philosophy of Science: Contemporary Readings*. New York: Routledge, 2002. This anthology works particularly well with Rosenberg's textbook (see below). Like the Rosenberg text, it is organized rather differently than our course, but it does an especially nice job of finding fairly accessible writings that nevertheless touch on the crucial issues.

Boyd, Richard, Philip Gasper, and J. D. Trout, eds. *The Philosophy of Science*. Cambridge, MA: MIT Press, 1991. This anthology has the virtue of bringing in a good bit of material from the philosophy of particular sciences. As a result, its coverage of general philosophy of science is a bit spottier than that of the other anthologies listed here, but it still does quite a nice job.

Curd, Martin, and J. A. Cover, eds. *Philosophy of Science: The Central Issues*. New York: W.W. Norton & Co., 1998. This is the anthology I use in my courses. I find it shockingly light on coverage of logical positivism, but the readings are, for the most part, otherwise well chosen, and this anthology includes extensive and user-friendly commentary from the editors. Given that even introductory anthologies in the philosophy of science consist of pretty difficult material, the commentary is especially useful.

Hitchcock, Christopher, ed. *Contemporary Debates in Philosophy of Science*. Malden, MA: Blackwell Publishing, 2004. Unlike the other books listed in this section, this is not really a comprehensive anthology, though it does cover a reasonable range of issues. It is organized as a series of debates, and it can be illuminating to see professional philosophers engaged in direct disagreement. Most of the contributions are both accessible and lively.

Essential Reading: General Textbooks

Godfrey-Smith, Peter. *Theory and Reality: An Introduction to the Philosophy of Science*. Chicago: University of Chicago Press, 2003. This is my favorite introductory text. Godfrey-Smith's sense of how to organize this material accords with mine, and his writing is clear and accessible. Because our course covers more material and does so in more depth than this book does, the text might appear a bit slight after working through our course, but it would make a nice accompaniment to it.

Hung, Edwin. *The Nature of Science: Problems and Perspectives.* Belmont, CA: Wadsworth Publishing, 1997. A thorough and accessible textbook that gives special attention to the study of patterns of reasoning that operate in science. It covers a lot of ground twice (once topically and once historically), which can be illuminating if not terribly efficient.

Rosenberg, Alex. *Philosophy of Science: A Contemporary Introduction.* London: Routledge, 2000. A nice introduction that simply approaches the material in a different order than I do. For this reason, this might prove a particularly useful book to some who have heard this course; it provides a different way of seeing how this material hangs together. Rosenberg focuses somewhat more narrowly than do the other authors considered here.

Essential Reading: Particular Topics

Ayer, Alfred Jules. *Language, Truth and Logic*, 2nd ed. New York: Dover Publications, 1952. The great positivist manifesto, at least in the English language. This work is enlivened by confidence and anti-metaphysical fervor. It presents positivism as a philosophical program, touching on philosophy of language, metaphysics, and ethics, as well as philosophy of science.

Berkeley, George. *Three Dialogues between Hylas and Philonous.* New York: Oxford University Press, 1998. There are many editions of this wonderful book, and just about any of them will do. See whether you can fare better than poor Hylas does as Philonous (speaking for Berkeley) attacks the idea that matter is a useful, a necessary, or even an intelligible concept.

Bird, Alexander. *Thomas Kuhn.* Princeton: Princeton University Press, 2000. Kuhn's ideas are of enormous philosophical importance, but he wasn't a philosopher; thus, it's quite useful to have an accessible, book-length treatment by a philosopher. For our purposes, one especially nice feature of this book is its emphasis on the positions Kuhn shares with his positivist predecessors.

Feyerabend, Paul. *Against Method*, 3rd ed. London: Verso, 1993. A provocative polemic against the idea of science as a rule-governed enterprise. Feyerabend celebrates the creative and anti-authoritarian side of science. He pushes pluralism a bit further than most philosophers think it can go, but the book is refreshing and contains important lessons.

Greene, Brian. *The Elegant Universe: Superstrings, Hidden Dimensions, and the Quest for the Ultimate Theory*, 2nd ed. New York: Vintage Books, 2003. An astonishingly accessible introduction to the strangeness that has dominated physics for the past century. Superstring theory is the main topic of the book, and I barely touch on that in this course, but along the way, Greene offers vivid and undemanding introductions to relativity and quantum mechanics, both of which make appearances in our course.

Hacking, Ian. *The Emergence of Probability*. Cambridge: Cambridge University Press, 1975. A fascinating story about how and why the central mathematical ideas of probability and statistics emerged in Europe in the 17th century. The book is mathematically undemanding and sheds light on economics, theology, and gambling, as well as history and philosophy.

Hume, David. *A Treatise of Human Nature*. New York: Oxford University Press, 2000. Hume's most important philosophical work, originally published while he was in his 20s. From the standpoint of our course, it is Hume's unwavering commitment to empiricism that matters most, but his unrepentant naturalism is of equal philosophical importance. This edition has some valuable editorial material, but other editions are more than adequate to the purposes of our course.

Kuhn, Thomas S. *The Structure of Scientific Revolutions*, 3rd ed., enlarged. Chicago: University of Chicago Press, 1996. The influence of this book can be overestimated, but it isn't easy to do so. *Structure* is often maddeningly unclear if considered as philosophy (rather an unfair standard to apply to a history book), but it decisively established the philosophical importance of the history of science and revolutionized the study of science from several disciplinary perspectives.

Sterelny, Kim, and Paul E. Griffiths. *Sex and Death: An Introduction to Philosophy of Biology*. Chicago: University of Chicago Press, 1999. How good a title is that? A fun and thorough introduction to the philosophy of biology, very generously laced with examples. I draw on only one chapter of it in this course, but I recommend the book as a whole. Readers should be warned, however, that *Sex and Death* is probably less even-handed and uncontroversial than some people expect introductory texts to be.

Supplementary Reading

Brody, Baruch, and Richard E. Grandy, eds. *Readings in the Philosophy of Science*, 2nd ed. Englewood Cliffs, NJ: Prentice Hall, 1989. This very useful anthology (full disclosure: I worked with Brody and Grandy when I was an undergraduate) has recently gone out of print, but as of this writing is easily and cheaply available through such outlets as amazon.com. It remains a convenient way to acquire some classic articles.

Carroll, John W., ed. *Readings on Laws of Nature*. Pittsburgh: University of Pittsburgh Press, 2004. Another case requiring full disclosure: Carroll is a colleague of mine at N.C. State. But I'm not alone in my opinion that he knows as much about laws of nature as anybody alive. As is to be expected in an anthology on a relatively specialized topic in the philosophy of science, the water gets a bit deep here, but this is as wide-ranging and accessible a collection of essays on laws of nature as one could ask for.

Clark, Peter, and Katherine Hawley, eds. *Philosophy of Science Today*. Oxford: Clarendon Press, 2000. An expanded version of the 50th-anniversary issue of the *British Journal for the Philosophy of Science*, this collection is aimed at professional philosophers, but the essays are surveys and some of them are fairly accessible. The collection as a whole provides a nice summary of the current state of the field.

Greenwood, John D., ed. *The Future of Folk Psychology: Intentionality and Cognitive Science*. Cambridge: Cambridge University Press, 1991. This volume stands at the intersection of philosophy of science and philosophy of mind. Given that we haven't learned much philosophy of mind in this course, some of the essays collected here are forbidding. But the philosophy of mind literature hasn't gotten any easier since this collection came out, and this remains a convenient place from which to survey debates about the relationships between folk psychology and scientific psychology.

Hacking, Ian. *The Taming of Chance*. Cambridge: Cambridge University Press, 1990. A follow-up to *The Emergence of Probability*, this book continues Hacking's engaging and interdisciplinary story through the 19th century. This volume tells the story of how chance came to make the world seem orderly.

Kitcher, Philip. *Science, Truth, and Democracy*. Oxford: Oxford University Press, 2001. This seems to me an important book. It begins by tackling issues about truth and reality, proceeds through some difficult issues about objectivity and interests, and culminates in a sure-to-be-influential proposal concerning the proper function of science in a democracy.

Kornblith, Hillary, ed. *Naturalizing Epistemology*, 2nd ed. Cambridge, MA: MIT Press, 1994. An influential and reasonably accessible collection of papers marking the "naturalistic turn" in epistemology and philosophy of science. Some of the essays are examples of naturalized epistemology, while others confront the philosophical issues about circularity and normativity that are highlighted in our course.

Ladyman, James. *Understanding Philosophy of Science*. London: Routledge, 2002. This is another nice introductory textbook, and it is especially generous in its explanations of the philosophical problems surrounding inductive inference. This book offers a somewhat narrow but impressively clear and helpful introduction to our field.

Larvor, Brendan. *Lakatos: An Introduction*. London: Routledge, 1998. I confess to feeling a bit guilty about including this work, rather than something by Lakatos himself, on the reading list. But Larvor is easier to read than is his subject, and the reader gets the added benefit of having Lakatos's marvelous work in the philosophy of mathematics presented, along with his contributions to the philosophy of science.

Nickles, Thomas, ed. *Thomas Kuhn*. Cambridge: Cambridge University Press, 2003. Kuhn's work has elicited an astonishing range of reactions, and it bears on a great many disciplines. This collection of essays by 10 different thinkers on 10 different topics is a good way to take the measure of Kuhn's legacy.

Pennock, Robert, ed. *Intelligent Design Creationism and Its Critics: Philosophical, Theological, and Scientific Perspectives*. Cambridge, MA: MIT Press, 2001. A collection edited by an unabashed critic of intelligent-design creationism, but one that lets advocates of intelligent design speak for themselves. Though Pennock is a philosopher of science, the collection ranges across history, politics, law, biology, theology, and education.

Shapin, Steven, and Simon Schaffer. *Leviathan and the Air-Pump: Hobbes, Boyle, and the Experimental Life*. Princeton: Princeton University Press, 1985. Like many philosophers of science, I have some serious qualms about the notions of truth, reality, and construction at work in much recent sociology of science. But this fascinating story about the history and politics behind the rise of the experimental method more than compensates the reader for any lingering philosophical irritations.

Sklar, Lawrence. *Philosophy of Physics*. Boulder, CO: Westview Press, 1992. This one is not exactly light reading, but it offers a clear and effective presentation of the deep philosophical puzzles that arise about space and time, quantum mechanics, and the role of probability in physics. Sklar does a lovely job of using the philosophy and the physics to illuminate each other.

Soames, Scott. *Philosophical Analysis in the Twentieth Century*, 2 vols. Princeton: Princeton University Press, 2005. This enormously useful survey centers more on metaphysics and the philosophy of language than on the philosophy of science, but this is a terrific way to learn about the developments in philosophy of language that made logical positivism possible. The later parts of the story illuminate Quinean holism and the theory of reference that helped make way for the resurgence of scientific realism.

Sober, Elliott. *Philosophy of Biology*, 2nd ed. Boulder, CO: Westview Press, 2000. A straightforward and sophisticated introductory text by one of the leading contemporary philosophers of biology.

Woolhouse, R. S. *The Empiricists*. New York: Oxford University Press, 1988. A brisk, reliable survey, not just of Locke, Berkeley, and Hume, but of the empiricist ideas at work in such predecessors as Bacon and Hobbes. This book is very friendly to beginners in philosophy.

Reference Works

Machamer, Peter, and Michael Silberstein, eds. *The Blackwell Guide to the Philosophy of Science*. Malden, MA: Blackwell Publishers, 2002. This is a very handy book. It consists of article-length essays, most of them about as accessible as their subject matters permit, that survey a topic or problem and offer suggestions about future developments. The volume touches on classic problems, such as explanation, but also on less common topics, such as metaphor and analogy in science.

Newton-Smith, W. H., ed. *A Companion to the Philosophy of Science*. Malden, MA: Blackwell Publishers, 2000. The *Companion* works differently than the *Guide* put out by the same publisher. This book is even more useful than its cousin. It offers short essays on dozens of topics and figures. Each entry provides the bare essentials concerning its subject, and most of the entries are quite accessible indeed.

Internet Resources

Philosophy of Science Resources. http://pegasus.cc.ucf.edu/~janzb/science. A terrific source of information, this page contains a great many links and construes the philosophy of science broadly.

Philosophy of Science Undergraduate Research Module at N.C. State. http://www.ncsu.edu/project/ungradreshhmi/evaluationModule/login.php. This mini-course was prepared by three of my colleagues at N.C. State (before I joined the department). The module covers demarcation, confirmation, explanation, and reduction. The writing is clear and helpful. The site requires users to obtain a password and to log in, but the site is free and requires no membership. The log-in enables granting agencies to track how many people use the module.

Science Timeline. http://www.sciencetimeline.net. An impressively detailed timeline that includes a surprising number of philosophical references and makes it easy to obtain more detailed information.

The Stanford Encyclopedia of Philosophy. http://plato.stanford.edu. This is a marvelous and growing peer-reviewed reference site for all of philosophy. Accordingly, not everything at the site bears on our course, and not every entry is accessible to non-specialists. But the site is easy to navigate and many of the entries are quite accessible.